ORTHOPTERAN ZOOLOGY

HOW TO KEEP GRASSHOPPERS, CRICKETS, AND KATYDIDS

Orin McMonigle

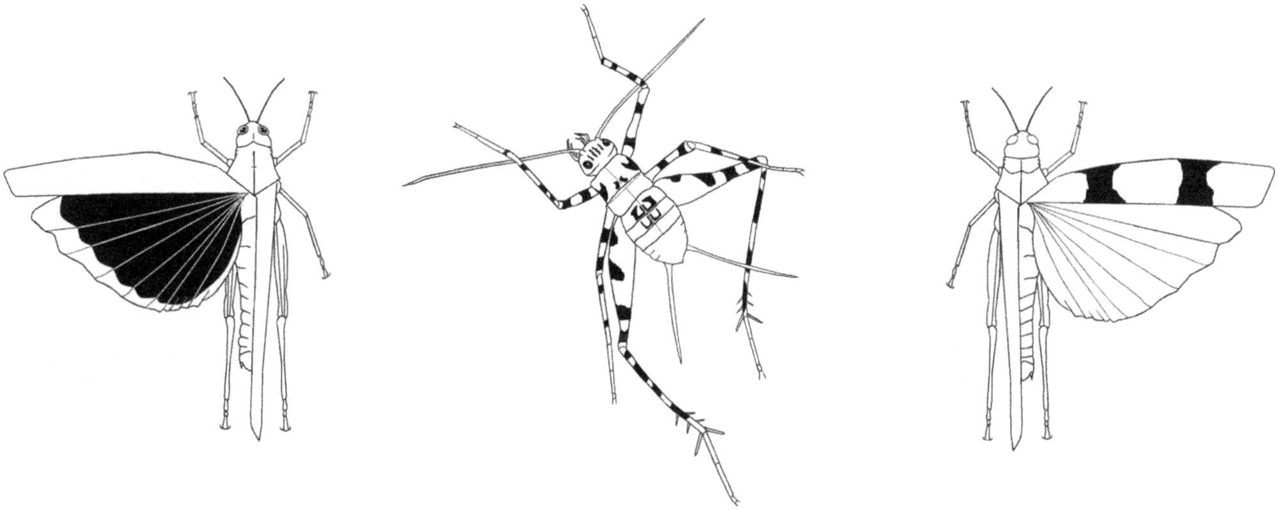

Coachwhip Publications
Greenville, Ohio

In Memory

In memory of Julius V. Turscanyi, who provided my first entomology text about orthoptera, and Barney Tomberlin, whose specimens' and offspring's photographs appear within.

Dedication

My thanks to Tristan Shanahan and Joshua Byrne for species information and photos, as well as to Ondrej Hes and Tom Larsen for providing photos and details for showy species commonly kept in Europe. Special appreciation is extended to Annie Lastar and Peter Clausen for editorial review, Steven Barney for editorial assistance, and Jayden T. Herrera, Hercules D. Hernandez, and Jesse Greene for specimens. I thank my wife Sylvia for support, my daughter Kree for searching Japanese pet shops and finding bell crickets and supplies to photograph, Gwynevere for bringing me grasshoppers from the field, and my son Jax who will do so one day. I thank God for providing the resources to put this book together and for creating these curious, big-eyed animals.

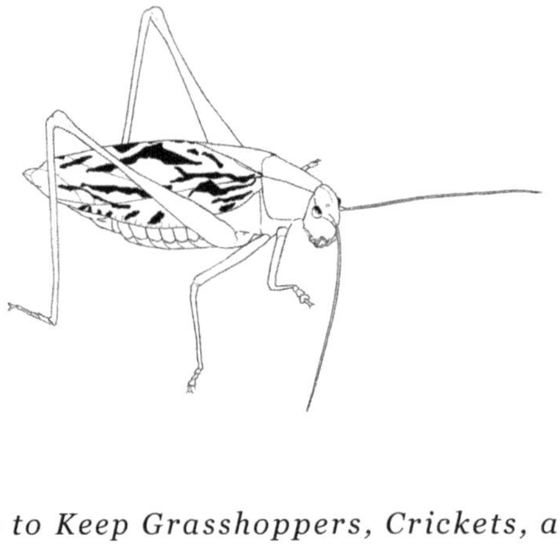

Orthopteran Zoology: How to Keep Grasshoppers, Crickets, and Katydids

Coachwhip Publications // CoachwhipBooks.com

Back cover images: Eastern lubber, *Romalea microptera* (CC-BY Judy Gallagher); Painted grasshopper, *Dactylotum bicolor bicolor* (CC-BY-SA Pavel Kirillov)

ISBN 1-61646-516-6
ISBN-13 978-1-61646-516-2

CONTENTS

Some katydid nymphs from Old and New World rainforests have upright wing pads that break up the body outline and aid in the crypsis of vegetation. Monte Verde, Costa Rica © Ondřej Hes

INTRODUCTION

There are numerous books containing a section on breeding house crickets as food for reptiles, arachnids, and other insectivorous pets. There are fewer texts dedicated to orthopteran identification lists or biological discussions. The rarest books have short sections, a few paragraphs to a few pages, devoted to cricket and grasshopper husbandry for display specimens (Simon 1975; Dunn 1993; Bruins 1999). Orthoptera husbandry and display comprises the first chapter in *Invertebrates for Exhibition* (McMonigle 2011). Unfortunately, none of these are primarily oriented to the discussion of orthopteran husbandry requirements—it is just a passing mention. *Orthopteran Zoology* is the first full-length book documenting the rearing and breeding parameters for these personable creatures. This text is written to aid keepers in the quest to learn more about the biology and husbandry requirements to successfully maintain orthopterans in artificial habitats.

Though books are lacking, there have been many fantastic articles devoted to husbandry—most often discussing one species' life cycle or unique data for hatching or survival. There have been dozens of articles in arthropod husbandry magazines including *Exotiske Insekter* (Danish), *Arthropoda* (German), *Invertebrates-Magazine* (English), and *Terrario* (French). Other magazines like *Mundo ArtróPodo* (Spanish), *Sklípkan* (Czech and English) and *The Bulletin of the Amateur Entomologists' Society* (English) occasionally have covered biology or specimen observation records in nature, but not husbandry. Oddly, there are nearly a

Invertebrates-Magazine, Dec. 2005
Neobarrettia spinosa cover

dozen *AES* issues with an orthopteran as the featured cover, but without accompanying articles (with only one- to three-sentence descriptions of the photographs). Orthopteran husbandry discussions can also be found on internet forums and pages, though posted questions often remain unanswered for months or years due to the rarity of the subject. Husbandry data can also be found in a small number of journal articles and difficult-to-find data sheets produced by zoo keepers. This book is an attempt to bring these various sources of data and rearing experiences together in one format and language on a permanent medium.

Invertebrates-Magazine, Dec. 2006
Romalea cover

1996 *Bulletin of the Amateur Entomologists' Society* with *Omocestus rufipes* woodland grasshopper cover

Grasshoppers, crickets, and katydids are fantastic pets and showy animals commonly seen on display at insect zoos. This group includes some of the largest and most colorful insects on earth. Most species are easy for insect zoo curators to rear if certain parameters are maintained (though some basic mistakes can doom a culture). Most orthopterans are excellent for exhibits because they display interesting behaviors, bright colors, and large size. Large, bulbous, compound eyes offer the illusion of great intelligence or intent in their actions. They are harmless to humans, although most will kick with spiny legs, spit unpleasant fluids, or bite with sharp jaws to avoid being eaten. Even the slow, clumsy lubbers can jump and are generally not used for petting zoos or handling displays. *Orthopteran Zoology* is written to provide enthusiasts, keepers, and beginning curators with the basic knowledge and understanding required to keep these creatures alive and reproducing in artificial habitats.

(Right) *Invertebrates-Magazine*, Dec. 2007 *Taeniopoda eques* cover

Katydid in Endau Rompin National Park, Malaysia © Ondřej Hes

Dead-leaf grasshopper (*Chorotypus* sp.)
Malaysia (CC-BY-SA Bernard Dupont)

The wart-biter, *Decticus verrucivorus,* ovipositing female
Belgium (CC-BY Frank Vassen)

Moncheca pretiosa
Mexico
(CC-BY-SA Pavel Kirillov)

Bush katydid nymph, *Scudderia* sp.
Virginia
(CC-BY Judy Gallagher)

Bradyporus macrogaster
Andros, Greece
Length 5 cm to 5.5 cm
(CC-BY Miltos Gikas)

Adult *Tropidacris collaris* kept in Denmark, from Netherlands vendor, 2002 © Tom Larsen

Schistocerca nitens is one of many related species in the New World, whereas in the Old World there is only one widespread species from this genus.

BIOLOGY

Classification: Order Orthoptera

Suborder Ensifera

Superfamily Grylloidea—
True crickets including mole crickets
Superfamily Hagloidea—Grigs
Superfamily Rhaphidophoroidea—
Cave crickets or cave wetas
Superfamily Schizodactyloidea—
Splay-footed crickets
Superfamily Stenopelmatoidea—
King crickets (wetas, Jerusalem crickets)
Superfamily Tettigoniidae—Katydids

Suborder Caelifera

Superfamily Acridoidea—
Acridid grasshoppers
SUPERFAMILY EUMASTACOIDEA—
Eumastacid grasshoppers
Superfamily Pneumoroidea—
Bladder grasshoppers
Superfamily Pyrgomorphoidea—
Pyrgomorph grasshoppers
Superfamily Tanaoceroidea—
Desert grasshoppers
Superfamily Tetrigoidea—
Pygmy grasshoppers
Superfamily Tridactyloidea—
Pygmy mole crickets
Superfamily Trigonopterygoidea—
Leaf grasshoppers

At more than 28,000 known species, the Orthoptera is among the greatest of the insect orders with incomplete metamorphosis. The variety of colors and shapes are endless. Some have long wings capable of taking them hundreds of miles across land and sea, while others have only armored tegmina (outer wings) useless for flying, or lack both sets of wings. They are among the largest insects in most habitats and are often the most numerous in terms of biomass. While some tropical species rank among the largest insects on earth, there are temperate adult grasshoppers and crickets with body lengths of barely 1/5″ (5 mm). Zoological display species are often 3-4″ (7.6-10.2 cm) and there are endless species in this range.

Brachytrupes grandidieri specimen from Madagascar next to a house cricket, both adult males

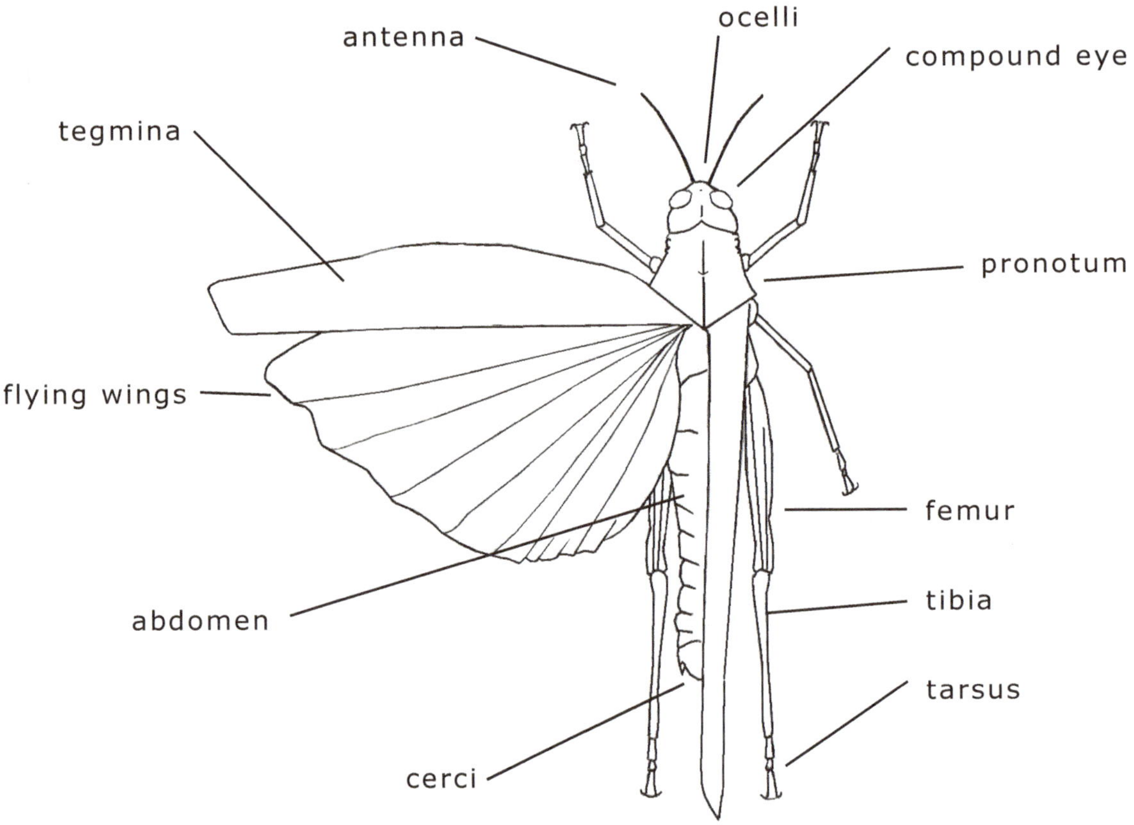

These grasshoppers (*Taeniopoda eques*) have hatched, wriggled to the surface and begun to molt.

Eggs hatch into nymphs that look very much like the adult and there is no pupal stage. In some ways this straightforward development makes them easier to care for than beetles and butterflies because the adults and young eat the same food and have similar requirements. (Though, pupation is not necessarily a difficult phase and does provide a break in the feeding regimen.) Still, lacking additional phases with unique requirements and changes in the feeding requirements does not mean orthopterans are easy to maintain without specific knowledge. Some species require specific temperature cycles, are sensitive to small changes in humidity, or are picky about food—or practical rearing methods are unknown and may seem impossible.

Avoiding its own predators, this huge predatory katydid is hunting at night. *Clonia* sp. female at Camp Satara, Kruger National Park, South Africa, 2013 © Ondřej Hes

Often, just getting the female to oviposit successfully or the eggs to hatch requires specific materials, moisture levels, or temperatures. Overcrowding presents a challenge to keeping most species, while others must be kept in individual enclosures.

Jax examines a lubber.

Within the Ensifera superfamilies, the most commonly reared are the (true) crickets, cave crickets, and katydids, while king crickets are wild-caught and maintained as novelty display items. Within the Caelifera, acridid grasshoppers are by far the most commonly kept, but a few pyrgomorphs and others have been cultured.

These *Acanthoplus discoidalis* were for sale at an insect fair in Malmoe, Sweden, 2007. This katydid genus is often sold under the name "armored crickets". © Tom Larsen

This katydid with a yellow underbelly is otherwise cryptic and difficult to see. It arches its abdomen to expose warning colors if a predator gets too close.

One of the largest grasshoppers in North America, *Brachystola magna,* circa Roswell, New Mexico

This katydid in Endau Rompin National Park in Malaysia is difficult to see on a tree trunk. © Ondřej Hes

Male *Ephippiger ephippiger,* Brac Island, Croatia: The sexes are similar in shape and color but the male lacks the long ovipositor. © Ondřej Hes

Female *Ephippiger ephippiger,* Brac Island, Croatia: Both sexes use the tiny wings to chirp when disturbed. Males also chirp to call females. © Ondřej Hes

REPRODUCTION

Orthopterans begin life in eggs that are deposited by the female; there are no live-bearing species currently known. In nature hatchlings usually never see their mother, even if she is still alive when they hatch. A fighting chance provided by a good hatching location and adequate nutrients in the yolk is all the motherly care they receive. Most eggs are elongate cylinders with rounded ends inserted into the ground singly or in clusters. The shell is usually leathery, pliable, and very tough. Eggs of certain katydids are compressed elongate discs adapted to fit within flattened leaves, stems, or bark. The female keeps the eggs inside her body until the right substrate and moisture is found. Temperature can also be a factor in where and when eggs are laid. Without the proper egg-laying site, some orthopteran females will never lay a single egg. Most eventually drop the eggs on the floor of the habitat if there is no adequate place to lay. The female may eat or chew on the eggs before they dry up and quickly die.

The cricket and katydid ovipositor are commonly a long sword or spear, while the female grasshopper has large valves at the end of a telescoping or rod-like abdomen. Most use their egg-laying mechanisms to bury eggs in soil or sand, but many of the katydids lay their eggs on or in vegetation. The shape of the female's ovipositor commonly reveals the shape of the egg and possibly where it is laid. With such an immense and varied group of insects there are bound to be exceptions to any rule of thumb. Orthopterans which lack ovipositors, like mole crickets, lay eggs in underground chambers. Mole crickets are known to look after eggs and early instar nymphs like earwig maternal care (Haes 1992). Jerusalem crickets have short ovipositors—one of the ways to tell *Stenopelmatus* from *Ammopelmatus* is by the length of the ovipositors—but egg-laying methods are uncertain.

Females and males are easy to differentiate for most species at an early age because the male lacks an ovipositor. Sexual differences are most notable at maturity, but after the first few instars, young specimens of most species are easy to sex at a glance. The developing wing pads can be proportionately much larger on young males and they should lack signs of a developing ovipositor. As they near maturity, males are often slenderer, and relatively smaller in proportion to the wings and abdomen of females of the same species. Many adult male grasshoppers are a fraction of the mass of the female,

(Top Left) The ovipositor of the female *Lesina intermedia* is a huge machete used to insert eggs in pseudostems.

(Top Right) Male *Lesina intermedia,* World of the Insect 2021

(Bottom Left) Although *Brachystola magna* males and females are similar in length, they are easy to tell apart because of the male's oversized back legs and the female's wide pronotum and triangular abdomen.

In this *Lesina intermedia* mating pair, the male hangs below. Mating took less than an hour.

including common *Melanoplus* spp. field grasshoppers and nearly all the stick hopper proscopid species.

One obvious difference between males and females that only shows up at maturity is the unique shape of the male's outer wings or tegmina. For most katydids and true crickets whose singing is legendary, the male's wings have large, visible adaptations in shape to create and amplify sound. Sexual dimorphism ensures only the males can sing, because the outer wings have these special structures. Female grasshoppers, crickets, and katydids are not always silent, but they lack strongly adapted wing structures and have limited or no singing ability. Male grasshopper wing adaptations for stridulation are difficult to see. Male and female grasshopper wings look similar, except in species where only the females have tiny or missing wings. Cave crickets and most king crickets (*Stenopelmatus sartorianus* is fully winged) do not have wings and are silent.

The primary use of the song is to attract females for mating. Without the male cricket and katydid, the loud symphony of insects on summer nights would be relatively quiet. Grasshopper stridulation is less noisy and often occurs during the day. Scientists have studied and documented

Comparison of female ovipositor valves (left) and male genital operculum (right), *Romalea microptera* adults

Mating pair *Ammopelmatus* sp. from Long Beach, California © Tristan Shanahan

the songs of all the common orthopterans with recording equipment. Even variation in songs have been studied for widespread species. The study of songs is an interest in and of itself without requiring one to know the animal's biology or even appearance. Sonograms that convert the song to a visual graph are easily located for most described species. Each species has its own unique call and most have a few calls for different uses. The calls of a single species within its geographic range rarely change in form, but they do vary in speed based on temperature. The most well-known case is when the speed of the song can be used to gauge the ambient temperature. In the lab or in the home, singing can be a blessing or a curse. Like any song, some people find joy where others find nuisance. Of course, the female of

First come, first served. Proscopid stick grasshoppers males are waiting on a female to finish laying her eggs (most of her abdomen is buried). © Tom Larsen

View from above, *Romalea microptera* oviposition

Side view of *Romalea microptera* ootheca formation

Romalea microptera ova removed and separated from the egg pod can still hatch if never dry and not re-buried. These are several months old.

Ammopelmatus pictus female with hardened spermatophore © Tristan Shanahan

Cricket eggs are as fragile as they look.

Neobarrettia spinosa, one-year-old eggs

the same species is expected to always find joy in the male's call or she will not have offspring. In orthopteran mating the female is usually on top and the male has the beautiful voice.

Females, as is often the case in the animal kingdom, expend huge amounts of energy and proportional weight in reproduction. It is not rare for a lubber grasshopper ootheca to be a quarter the size of the female that laid it. What is a little different is the amount of investment by the male—especially some katydids who invest up to thirty percent of their body

Ovipositor of *Ammopelmatus pictus* female from West Stanislaus County, California © Tristan Shanahan

weight in a single spermatophore. (The unimaginable equivalent would be 60 lbs. for a normal 200 lb. human male.) The reason is two-fold. First, most orthopteran females only require a single mating to produce fertile eggs for the rest of their egg-laying lifespan. They have a chamber or paired chambers (spermathecae) where the spermatophore is stored and used to fertilize eggs as they are laid. Second, the spermatophore is often designed to plug the entry and the female chews off the excess so it does not get in the way as she moves about. The excess surrounding material, known as the spermatophylax, provides her crucial nutrients in her task of producing eggs and allows the male to assist in the process.

Neobarrettia spinosa female with a fresh spermatophylax after mating © Joshua Byrnes

Snowy tree cricket, *Oecanthus fultoni,* courtship, Brunswick, Ohio, 2021

A male *Homoeogryllus xanthographus* sings. Wings of crickets are lifted and opened during singing so it is easy to see which specimens are chirping. © Ondřej Hes

This true katydid female is trying to find a good spot to insert eggs with its short ovipositor. It was found as a nymph in Medina County, Ohio.

Ammopelmatus pictus ovum
© Tristan Shanahan

Scudderia furcata, male genitalia

Scudderia furcata, female genitalia

There is a significant size difference between adult male and female stick grasshoppers (*Stiphra* sp.). Males are the smaller green specimens.

Camel cricket with spermatophylax, Malaysia (CC-BY-SA Pavel Kirillov)

Mating pair of *Monistria maculicornis,* Australia (CC-BY Jean and Fred Hort)

Schistocerca gregaria in Croatia © Ondřej Hes

RELATIONSHIP WITH MAN

The most famous Orthoptera, since biblical times, have been the locusts. Massive swarms decimated whole countries and caused starvation across Europe, Africa, the Middle East, and western Asia as recently as 2021. Clouds of locusts blot out the sun and fly at a speed of about 18 kph (11.25 mph) (Stanek 1969). *Shistocerca gregaria* holds the record for migrating over 2,800 miles (4,500 kilometers) from Africa to South America (Rainey 1989). The Old World desert locust is not normally found in South America, where its counterpart *Shistocerca cancellata* reigns.

The industrious human could simply collect the grasshoppers and eat them instead of worrying about lost crops, but a common aversion to eating insects often prevents this solution. Unlike crops, the grasshoppers must be collected and killed before they fly off. They also must be dried quickly or the carcasses rot in a few days and become inedible. North America no longer sees equivalent grasshopper swarms. The Rocky Mountain locust (*Melanoplus spretus*) plagued western North America before becoming extinct around 1900. Swarms of Mormon crickets (a type of katydid) decimated the crops of early settlers in the southwestern United States and still swarm today. Periodical cicadas, barely related to Orthoptera and relatively harmless to crops, are the only 'locusts' with which most people in North America are familiar.

Locusts are the migrating forms of grasshoppers—they exhibit different colors and have longer wings than is seen in the normal growth pattern. Migrating and solitary forms of *S. gregaria* occur primarily dependent on the density of the locust population. The locust form is easily created in the lab by rearing small nymphs in crowded conditions. In nature, overcrowding is attributed to weather conditions that are perfect for optimum hatching and growth.

Countless billions have been spent on monitoring, studying, and formulating chemical poisons and physical barriers with uncertain evidence of success. The documentary *Locusts: War Without End* (1981) shows government officials working with radar, satellite imagery, and predictive computer models to try to combat, forecast, and prevent the largest swarms of *S. gregaria*. The response to swarms is massive doses of poison broadcast by airplanes. Officials explained this could help reduce growth of the swarms, but it is primarily responsive and kills non-target

Locust swarm in Israel (CC-BY-SA Niv Singer)

Locust swarm in Madagascar (CC-BY-SA Hotel Kaesong)

organisms. The imbalance of predator and parasite kills could cause rather than prevent future plagues. Much of the documentary tried to explain how massive swarms started in 1977 were missed by control officers when preventative measures might have been effective. Swarms and plagues are often unpredictable, and weather related, though not weather-created cycles. Cycles of plague are tied to weather cycles because grasshoppers do better in dry, hot years of drought. The droughts themselves are tied to the 22-year cycle of sun spots (Mitchell et al. 1979). Controlling plagues may be no easier than controlling the weather, but fear and a feeling of powerlessness are great drivers of human behavior. Many of the measures man takes gives the feeling of control whether or not they really work. Workers can take credit and be given credit for good years either way. There have been no significant swarms in North America since the 1800s—no entomologists credit this to a specific effort, though it was probably incidental to human activities.

Cricket suckers are sold as novelty items in many North American museum gift shops.

Plastic grasshoppers are common, but plastic crickets like this weta-looking monstrosity are unusual.

For human consumption, crickets and grasshoppers are canned and readily available in a few countries. They have become increasingly popular in the health food niche in western cultures in recent years. As worldwide population grows, different sources of human food will have to be investigated, but eating Orthoptera may just be a fad. Even in countries with a long history of insect cuisine, such as Thailand where fried hoppers on sticks are found at certain food stands in major cities, eating insects is only a novelty. The general population has never eaten one. Likewise, many museums in North America sell suckers with a cricket inside, a curious gag gift that is more often tossed out than consumed. Most crickets reared in bulk by humans are used as food for insectivorous pets.

Plastic grasshopper and cricket toys exist in uncountable quantities, but usually in a bag of plastic bugs sold for uncertain uses. They are rarely anatomically correct, so a green rubber toy might be

Jiminy Cricket (1940)
at Disney World, 2021

Refrigerator magnets in gift shop, at
Dr. Crawley's Insectorium, Kentucky

Romalea microptera on display in Washington D.C., 2011

Specimen on office wall. I was given this as a gift from Bangladesh many years ago and thought it was a pyrgomorph grasshopper without ever really looking at it. It is a katydid, *Sanaa imperialis*.

This museum display shows predatory katydid specimens. (An insectarium includes live specimens, an insectorium only dead ones.) Dr. Crawley's Insectorium, Kentucky

a katydid or a grasshopper. Most are a single lump of molded plastic or rubber, with little or no painting, but there are toys formed with dozens of parts or added electronics. One of the most famous toys is the 1984 Transformers' Insecticon™ grasshopper, 'Kickback,' advertised in numerous appearances on the animated series from 1984-86. Another famous grasshopper inspiring a large line of associated toys was the evil character 'Hopper' in the 1998 CGI movie *A Bug's Life*. One of the 'Hopper' toys was electronic and as big as a house cat. 'Cri-kee' from the animated *Mulan* movie inspired a few hard-to-find toys. 'Jiminy Cricket' from *Pinocchio* (1940) might be the most famous and longest-running orthopteran toy inspiration, but without much of a toy presence.

Insect zoos in the United States have displayed grasshoppers, specifically *Romalea microptera*, since at least 1978.

The earliest exhibit was at *World of the Insect* in Cincinnati, Ohio, followed by the Smithsonian insect zoo in Washington, D.C. The massive South American *Tropidacris* and Southeast Asian *Aularches* became popular in European insect zoos in the 1990s. In the 2000s stick grasshoppers gained some prevalence in displays. Grasshopper displays usually hold a large number of specimens of one species. Many insect zoos today are likely to host two or three displays.

Katydids have been widely displayed across various insect zoos since the early 1990s, especially cone-headed species from Central America and dragon-headed katydids from Southeast Asia. Predatory species, even those that consume some vegetable matter, are normally kept one to a display. Captive-bred pink and orange color variants of the native (normally green) oblong-winged katydid, *Amblycorypha oblongifolia*, have been a large exhibit at the Audubon Insectarium in New Orleans, Louisiana, since shortly after its opening in 2008. Giant Malayan leaf katydids (*Arachnacris* sp.) have become regular display animals at insect zoos worldwide over the last decade and are some of the most massive and impressive katydids on earth. They are kept in relatively close quarters since they normally do not harm each other. As with grasshoppers, it is unlikely to find an institution displaying more than three species of katydid.

Cubed cricket diet is available at most North American pet stores.

Transformers' Insecticon™ Kickback, 1986

Other than as food for animals in arachnid and centipede displays, crickets have seen little action in zoo displays. I have seen a few temporary Jerusalem cricket displays, but they are not true crickets, and it is difficult to make a good display of a solitary burrower. There are some huge tropical crickets, but they are not available from insect farms. They are mostly nocturnal, making display a little trickier. Despite a lack of presence in institutional displays, when it comes to captive-breeding history and quantity, crickets are man's most beloved orthopteran.

A thousand years before feeder crickets, various species were cared for by humans for their beloved songs, not for impressive display size or appearance. Wild crickets have been collected for their musical abilities in China and Japan for at least 1300 years. Cages built and used to house singing crickets in China date back at least to 700 A.D. The bell cricket (suzumushi) is a popular pet in Japan and has been bred and reared in artificial habitats there for over 250 years (Larsen 2019). Cages, pelleted and jelly foods, shelters, media, and complete kits including eggs are available at department stores and general pet shops across Japan. Even 'cricket' water is packaged and sold for their consumption. The bell cricket is a handsome black creature with white antennae and cerci, but a large one is barely an inch long.

Since a few orthopterans are the most infamous pests of agriculture, all exotic species, except for established adventives like the house and banded cricket, are restricted by government agencies in North America. The beautiful, large, exotic species are only maintained by keepers in insect zoos with approved containment facilities. In Japan, the native bell cricket is the only species offered in pet shops and mail order. In the rest of Asia and South America orthopteran husbandry is mostly academic if not restricted to insect farms providing specimens to insect zoos, the dried specimen trade, and (less often) enthusiasts in Europe. In many European countries where the migratory locusts sometimes dally, amateur entomologists have been able to acquire captive-produced young of beautiful species from the worldwide tropics since at least the 1980s. Many are kept for a few generations and traded but never seem to have the staying power of the much more commonly kept phasmids (of which some current cultures date back to the 1960s). In Europe it is more likely to see native species off limits—mostly local populations threatened by habitat loss.

Shown is a beach grasshopper, *Trimerotropis maritima,* adult from the southern shore of Lake Erie. This species is found widely, so wholesale habitat loss (from beachfront housing and recreational beaches) has not affected it as greatly as other dune species.

Even for amateur entomologists restricted to maintenance of only native

Shown are bell crickets for sale in Tama City, Japan. This species is commonly on sale in August and September at general pet shops and home (DIY) stores, rather than at beetle shops. © Kree McMonigle

Packaged wood, water, and jellies for bell crickets at the same shop © Kree McMonigle

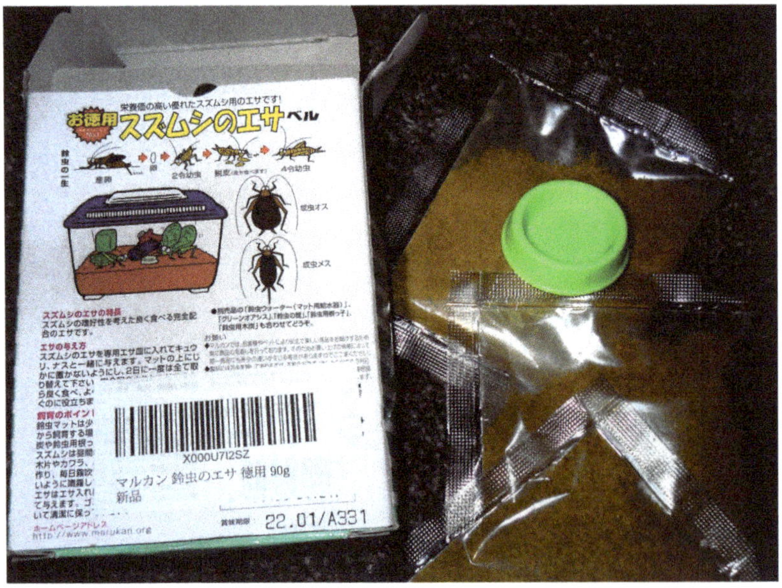

Packaged Suzumushi food aquired from a Japanese vendor through Ebay

The beach grasshopper *Trimerotropis maritima* looks nearly identical to *Trimerotropis huroniana,* which has been decimated by habitat destruction and is protected by Canadian law. It occurs alongside *T. maritima* on northern Lake Erie shores and is easily differentiated by markings on the inside of the rear femora.

Shown is a subadult female *Dissosteira carolina* found on a Lake Erie dune, though this specimen was probably only found there because of nearby gravel roads and paths.

species, there is a lot of room to learn and grow. Although many of the massive and colorful species live in the tropics, even cold areas (devoid of most other interesting insect life) host orthopterans in great variety. I am from one of the colder areas of North America and over the decades I have been amazed at just how many different orthopterans there are, from pygmy grasshoppers to flying giants along the roadsides, birdwings, purple winter grasshoppers, huge *Melanoplus* females, handsome trigs, snowy tree crickets, fantastic cave crickets, and a variety of different large, green, leaf-like katydids.

Natural sand dunes are rare because they have been cleared for housing and public beaches all along the shore. This is a small natural area on Lake Erie in Mentor, Ohio, on a utility property. *Trimerotropis* hang out around the edges of the sand dunes and are practically impossible to catch once they have wings. They fly away, expose yellow and bluish startle coloration, sometimes make a clacking sound, and land five to ten feet (1.5 to 3 meters) distant. The pursuer will always tire before the grasshopper. The pursued grasshopper eventually flies greater distances and knows not to land in the water.

Artificially recreated dune habitat with giant blue grass, west of recreational beach
Cedar Point, Sanduksy, Ohio

Trimerotropis maritima from artificially recreated dune habitat
Cedar Point, Sanduksy, Ohio

Screen cages usually offer good molting surfaces for large grasshoppers and katydids even though the surfaces are at right angles. This is a late instar nymph of *Taeniopoda eques*.

HUSBANDRY

Enclosures

The first and most important parameter for maintaining orthoptera is choosing or constructing a practical enclosure for both the species and quantity of specimens to be kept. Two widely-used display enclosures are glass terrariums (with screen lids) and full-screen cages. A utilitarian cage for anything can be made with a large plastic tub, some screen, and hot glue. Things to consider when choosing size and venting include ambient humidity and the species' humidity needs, surface texture for molting, size (both height and surface area), and the ability of the construction materials to contain its residents. Generally, only metal screens should be used for vents. Many species quickly chew holes through nylon window screen, slower through cloth screens like muslin.

Ventilation is the most important aspect and the easiest to control, but not always the easiest to adjust before problems occur. Ambient humidity and air flow are important to consider and change seasonally. If proper humidity is not maintained and the animals dry out too quickly, molting problems occur. Except for burrowers, orthopterans are damaged and killed by damp, stagnant air. At the least, a few large vents should be added to the enclosure. The screened surfaces are often used as molting purchases since large grasshoppers and katydids need to molt hanging from something the claws can hold onto. (Most crickets do not need to hang and can survive a molt on a flat surface, while mole crickets molt in an underground chamber.) Full-screen cages can be very successful in offering the greatest area of good molting surfaces, but are likely to result in unsuccessful molts if the ambient humidity is low. If the humidity is below 25%, misting a few times a day may solve the problem, but that may not be enough. If the humidity is greater than 60% the animals usually do not drink water because they get enough moisture from the food. In humid conditions with limited ventilation and airflow, misting will cause the frass to mold and create unhealthy conditions.

The size of the cage relates to the required molting dimensions for adults, as much as to how many specimens can fit. Few species can be kept well in anything with smaller dimensions than a 5-gallon (19 L) glass aquarium, but two or three times that size is often a minimum. An enclosure may be big enough to readily fit a few animals, but still not large enough

Cork bark is commonly sold for terrarium use. It works great for shelter and molting while giving the enclosure a natural appearance. These *Ceuthophilus gracilipes* nymphs are much lighter in color than adults.

Copiphora rhinoceros adult pair cohabitation at World of the Insect, 2021

to hold adequate containers or plants for egg laying. Overcrowding is a consideration for most species, but for katydids the maximum number is often one. Even vegetarian katydids are prone to snacking on cage mates that are molting. Species kept individually are practical to rear in smaller cages, specifically a tall deli-cup (32 oz. or 0.95 L—the same container is often used for grasshopper egg pod retrieval and storage).

Large grasshoppers and katydids require molting surfaces two to three times their body-length to molt well and emerge with perfectly formed wings. The surface should be easily gripped by the animal, which eliminates smoother surfaces for huge species. The opportunity to find a spot to molt without being knocked down by other animals is also necessary for survival. New keepers often lose specimens due to inadequate or improperly placed molting surfaces. Large branches with rough bark should be placed at a 70° to 80° angle from the substrate. Animals can accidentally molt too close to the bottom if branches are placed at a lower angle. Screen caging offers good molting surfaces, but perpendicular and horizontal surfaces do not work well for all species. When judging adequate molting surfaces through failures, do not forget that low ambient humidity can cause similar-looking damage.

A clear distance from the bottom or highest decoration to the top of the enclosure is important for terrestrial species so they cannot jump out as soon as the lid is lifted for feeding or cleaning. For arboreal species which hang out on the sides and top of the cage, opening the lid or door on a screen cage can result in leaps to freedom, smashed specimens, and unnecessary wounds to the caretaker. Cages for large, aggressive, or flighty species should have a hand-sized access door used during replacement of food plants, introduction of prey, or arrangement of decorations and hide spots. With big predatory katydids like *Neobarrettia* kept in smaller, individual enclosures, a foam-stoppered feeding port needs to be just large enough to introduce prey items (usually house crickets).

Substrate is not just for egg laying and maintaining humidity. This mole cricket nymph is digging through damp sand.

Water

Humidity and moisture are not the same thing when it comes to orthopteran husbandry parameters. Humidity is water vapor in the air and may be very stratified in artificial habitats. Humidity can be near saturation an inch above a water bowl and less than 20% near the lid or above dry substrate on the other side of the enclosure. Moisture is the normal term for liquid water, from tiny droplets in the food or substrate to a pool of water in a dish. Orthopterans generally need to be provided moisture (misted with a spray bottle once a day) if the ambient humidity is low, below 40%, or if the primary diet has a low moisture content (dry dog food or frozen leaves). If the primary diet has a high moisture content (various melons, lettuce leaves) extra ventilation may be needed to dry out the watery frass before it molds.

Most grasshoppers, crickets, and katydids do well with good access to moisture, but cannot survive high humidity without excellent air flow. If there is

Indian house crickets hanging out in packing carton shelter

Egg carton is used for shelter and molting for many true crickets and cave crickets. This is a group of *Ceuthophilus agassizii*. © Tristan Shanahan

slightly high humidity and limited airflow for days or weeks, mold will grow in the trachea and the animal suffocates and dies. Mold in the trachea is usually not visible with the naked eye because it is not possible to see inside the trachea of living animals. More common than mold, many orthopterans are killed rapidly from too much moisture and not enough ventilation because they suffocate. If specimens show signs of lethargy after being kept in a damp holding container, they seldom recover. Most arboreal and terrestrial orthopterans readily survive three days with no food or water but die after a few hours in a closed jar containing wet grass or damp paper towel. Subterranean orthopterans like mole crickets, cave crickets, and Jerusalem crickets are adapted to high humidity, can survive limited ventilation, and are more likely to desiccate from dry conditions.

In extreme cases all orthopterans, even desert species, will dehydrate and die. Under the care of a watchful keeper, less than adequate moisture is more likely to lead to unsuccessful molting. Too much evaporation and too little consumed water may cause the insects to become stuck in the old exoskeleton and die, develop crumpled wings, or improperly molt the trachea linings. Trachea linings appear as white strings sticking out of the spiracles following a molt if the animal was hydrated enough to molt, but not enough to molt correctly. Health problems caused by too little or too much humidity should be watched for and adjusted as needed.

Food

Orthopterans, especially grasshoppers and katydids, can often be reared entirely on fresh leaves of specific plants. (Even the most polyphagous species has a relatively limited range of leaves it can survive on.) The leaves of a narrow range of plants are best for a wide variety of species, much as with the phasmids. There is no need to offer a variety of acceptable leaves unless you do not know what they will eat. The use of one well-accepted food plant is usually most successful. Additional types of food like fresh fruit slices may be helpful, but have little impact on size or survival. Fresh blackberry or raspberry (*Rubus* spp., also known as bramble) or leatherleaf viburnum (*V. rhytidophyllum*) leaves can be used as singular foods for a wide range of grasshoppers and katydids (and some crickets like the tree crickets from the Subfamily Oecanthinae). Oak (*Quercus* spp.), privet (*Ligustrum* spp.), and rose (*Rosa* spp.) are eaten by many of the same species. Romaine lettuce is a favorite food of many grasshoppers that is also available year-round at grocery stores. When other leaves are readily available, romaine is often not the best choice because it requires extra effort and replacement to keep fresh. Withered and yellowing leaves result in reduced feeding and sometimes even death of the orthopteran. Usually, one type of plant is used. Food items can be changed during the life of grasshoppers in captivity and some polyphagous species hardly seem to notice, while others react poorly to alternating food plants.

Leaves of bramble, rose, viburnum, and privet can stay fresh for days or weeks using a well-planned water holder. The normal method is to cut a fresh branch and submerge the stem in a container of water to keep leaves fresh like cut flowers. A 16 oz. (0.47 L) deli-cup is sized well even for a small enclosure like a 5-gallon (19 L) terrarium. Sour cream and other food containers come in the same dimensions as deli-cups, but they are opaque. A clear container is best for monitoring water level and fouling. The deli-cup normally used for wonton soup is made of a thick, soft, semi-opaque plastic which usually holds up around the stems after multiple uses without cracking. The container lids become damaged and fall apart with use. Lids must be replaced before damage is excessive because small grasshoppers and katydids seek out relatively small gaps and drown.

Painted grasshopper, *Dactylotum bicolor*, feeding on a flower (*Hibiscus syriacus*).

Food offering of romaine lettuce, sweet potato, and apple for *Arachnacris corporalis*

Leaves with large stems like lettuce, cabbage, and cattail can be kept in a similar container without a lid. The container is filled with damp sand and the thick stems of the leaves are buried in the sand.

Alternative or supplemental vegetable matter provides nutrients a leaf diet cannot, but usually does not have a measurable impact on survival or growth. Still, sliced fruits and vegetables may aid survival if leaves are in limited supply or in poor shape. Flower petals from roses and rose of Sharon (*Hibiscus syriacus*) are seasonally available in quantity and relished by many species. *Hibiscus rosa-sinensis* (cultivated variety) is a great food plant and laying site for some katydids. Apple and pear slices and carrots are chewed on by many different orthopterans.

Terrestrial crickets are normally reared using chicken mash or various kibbled foods created for predatory animals—the most popularly employed and useful are dog or cat food. The main ingredient is usually corn but other meal or meat-based foods work similarly well. Gerbil or rabbit pellets and some fish foods are too low in fat and likely to be ignored barring starvation. Grasshoppers and katydids, whether mostly 'vegetarian' or predatory, usually chew on the same kibbled foods accepted by crickets. However, long and short-horn grasshoppers very rarely survive on kibbled foods as a singular source of sustenance.

Predatory katydids eat live insects and avoid leaves. Unlike some predators that lack the ability or desire to kill their

While this female *Ephippiger ephippiger,* Brac Island, Croatia, was photographed on bramble, it is a predator. Many predatory katydids consume phasmids that are easily cultured as feeders using bramble as food. Predatory katydids do not always adapt to climbing down to feed on terrestrial crickets. © Ondřej Hes

own, predatory katydids are dangerous to each other. Sometimes the 'vegetarian' katydid species, as well as crickets and grasshoppers, are also dangerous to cage mates. Usually, this predation takes place when other animals are dying or molting and cannot walk away. Cannibalism is usually determined first by the species, second by overcrowding, and third by available food or moisture. Wing biting is usually moisture or space related, but thorny foodplants may rarely cause similar looking damage.

Although limited information is available for orthopterans, tiny fresh leaves just opening from winter buds of European bramble (*Rubus fructicosus* species complex) in early spring (around April) are toxic for some of the big, popular phasmid species (*Phyllium, Extatosoma,* and *Heteropteryx*). Fresh spring leaves of wild and cultivated blackberries in North America are not harmful to *Melanoplus* or native katydids. Seasonality of foods may play a small role in choosing the food plant, at least in early spring, in Europe for captive eggs that hatch early.

Nymphal Development

Most orthopteran species molt a set number of times on the way to maturity. Small nymphs molt every few weeks or less, with the time between lengthening as the animal grows. The final molt takes the longest and usually results in wings for singing or flying. Adverse conditions cause them to take longer, mature smaller, or even reduce wing growth, rather than molting more times. Supernumerary molts (those beyond standard) are rare and likely more of a genetic defect than a coping mechanism. Grasshoppers commonly undertake the first molt within minutes of hatching. They hatch and wriggle to the surface by contracting and expanding the head and twisting the body. The legs of the hatchling are folded back against the body and are useless for locomotion. For grasshoppers, their five molts include the molt after hatchlings wriggle to the surface. Katydids, true crickets, and mole crickets usually molt eight times, sometimes more, and do not molt immediately after hatching.

Early nymphs of both genders look similar, but the shape of the body and ovipositor differentiate within a few molts. Males are slighter bodied and usually mature more quickly. Both genders usually require the same number of molts.

In captivity, the food offered to all stages from hatchling to adult is usually the same, with the exception being predatory species. The type of prey is specific to the size of the predator (as prey cannot be too large to tackle or too small to grasp).

The main differences in habitat requirements for different stages is found in moisture and molting surface needs. Small specimens usually require higher humidity or more regular misting than large nymphs or adults. Large specimens are more easily harmed than small animals if there is not enough ventilation, but even big specimens can require more regular misting for drinking if the humidity is low (especially below 25%). Large grasshoppers and katydids can usually molt off almost any surface (even smooth glass) in the first few instars as they hang by the euplantulae. At a certain weight and instar, developing nymphs require larger molting areas with a rough surface to hook the claws into. Grasshoppers,

Molting distance needed for a safe molt is illustrated by this image of a katydid in Val Estrella, Costa Rica. © Ondřej Hes katydids, and certain crickets must hang upside-down in order to molt properly. If they fall because of an incorrect surface, they become severely deformed and die.

Teratology and Mutations

An acridid grasshopper in Africa was found with a leg-like appendage growing in the place of an antenna (Kevan 1953), possibly a result of the same antennapedia gene responsible for this malformation in fruit flies. Other specimens have been found with two to four median eyes where there is normally one (Cockayne 1930). The most spectacular developmental oddities are the bilateral gynandromorphs that display sexual dimorphism in one specimen. Such specimens have been described for various grasshopper species (White 1968). Abnormal homeotic development is rare in captivity, but with a similar rate of occurrence in the wild. Across many specimens I have not reared one with an extra leg or eye, not even a tergite or sternite malformity. Segment malformations are well-known, but rare for the related phasmids and roaches and likely occur in similar ratios (less than one in a thousand). A keeper can rear thousands of specimens over decades without coming across any of these developmental oddities.

Some grasshoppers normally display variation in wing length in captive specimens maintained under the same conditions. Many orthopterans can reduce the energy for wing production and adult size, or transform from green to brown as a result of differences in rearing conditions (humidity, lighting, food moisture content, or crowding). Locusts grow longer wings in crowded conditions. Standard variations within a species or environmentally caused changes in shape or color are normal. Creatures that express recessive genes for color (not just color differences caused by environmental variables) are often called 'mutants.' Simple recessives are encountered within a few generations of line breeding if the genes are present in the population sample. There are various reports of 'albino' or pink katydids or purple grasshoppers in the literature over the last hundred years (Brock 1993; Sutton 1992 and 2004). However, oddly colored specimens are likely to be eaten by predators before they reproduce in nature and therefore remain a barely seen portion of the population. Many species naturally have pink or purple highlights, while the unusual specimens lack the basic green or brown coloration of the normal form. I am not aware of anyone concentrating pink mutations in captive grasshoppers, but albino forms of *Locusta migratoria* were isolated in early 2000s and have been

Acheta domesticus, hypermelanistic form, 2012

Pink katydids from W. S. Blatchley's "Orthoptera of Indiana" (1902), reprinted from Scudder's "Pink grasshoppers," *Entomological News* (1901)

Pink varieties: (Top) *Chrysochraon dispar*, Belgium (CC-BY Frank Vassen); (Left) Katydid, New York (Richard Whitby); (Right) Grasshopper, France (CC-BY Gail Hampshire)

An unusual katydid from Malaysia (CC-BY Zleng)

A pink grasshopper from the UK (CC-BY Tim Parkinson)

A yellow katydid from South Carolina (CC-BY Hunter Desportes)

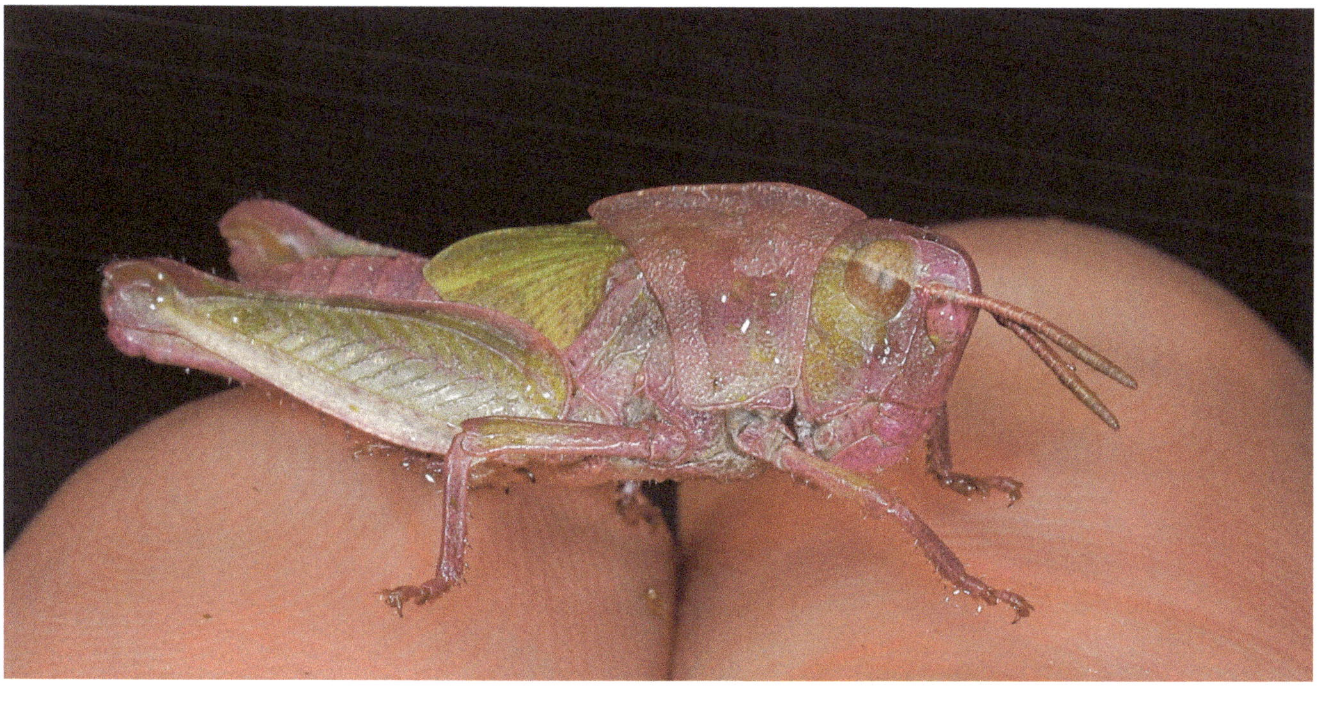

Some orthopterans have natural pink highlights. Northern green-striped grasshopper, *Chortophaga viridifasciata viridifasciata,* Maryland (CC-BY Judy Gallagher)

Cultivated pink *Amblycorypha oblongifolia* on display at the Audubon Insectarium in New Orleans, Lousiana, 2014

An unusual yellow-and-green variegated specimen of *Amblycorypha rhotundifolia*, found wild, 2010

Indian house cricket female with full length tegmina and functional flying wings versus standard female: this may be how this species increases its range without sacrificing the energy required for wing growth in all specimens. Such specimens can result from environmental cues, population genetic plasticity, or common defects in development.

Note the unusual congentital segment deformity of this Jerusalem cricket. *Ammopelmatus* sp. from Long Beach, California. © Tristan Shanahan

used for various experiments. Though no less beautiful than a ghost white locust, some effort has been made since 2008 to select for spectacularly colored pink, yellow, and orange katydids for zoological displays. The normally tan house cricket has been line bred for size, longevity, and hypermelanistic (black) adult coloration (McMonigle 2013).

Fertilization

For the vast majority of orthopterans, captive mating is straightforward and consists of simply keeping adult specimens of both sexes together in an adequate enclosure for a few hours or more. In general, males will wait two to three weeks after the molt to maturity before they are ready to mate. Females, depending on the species, may wait, or mate immediately while the exoskeleton is still soft and pale (teneral mating). There is usually a little bit of a courtship dance before mating with wing flapping and antennae touching. Fertility issues over generations have not been observed, though, other than a few cricket and grasshopper species, few have seen more than a few consecutive captive generations.

In the case of predatory katydids, some can be difficult to mate successfully because the female eats the male before he gets close enough to mate. In artificial habitats, predatory species are usually reared individually (at least after the first few molts). Mature pairs are later introduced in a large cage with branches and leafy plants for the male to hide. For the first few minutes to hours, they should be observed for aggression and separated when necessary (similar to mating mantids and tarantulas in captivity).

Though nearly all species require a male and female for reproduction, there are some rarities like the parthenogenetic *Saga pedo* for which males are unknown. *Saga pedo* is one of the world's largest predatory katydids, found from the Mediterranean to China. It has huge eggs, 11 mm long by 2.5 mm wide, and is a difficult-to-keep species (Hromadka 2000). While males are status quo for other *Saga* species, one must wonder if the very aggressive, predatory nature of the female was an important driver of this situation. There are some other rare cases of parthenogenetic orthopterans, and these probably did not eat all their mates. They include a grasshopper from Australia, a cave cricket from North America, and a true cricket from South America (Fernandez 2015).

Ovum Care

Egg-laying trays or containers are a standard in orthopteran husbandry. Any type of plastic container with a mouth 8 cm (>3″) or wider has been used by keepers. Some species have difficulty finding the trays so the entire bottom of the habitat must be lined with containers or filled with substrate for egg laying. Removable trays and containers may be chosen with a fitting lid to maintain moisture of the substrate mix during long gestation periods. The lid, of course, is added after the eggs are laid. Cricket trays can be shallow because females do not need a tray deeper than the ovipositor, which is rarely more than 2 cm (4/5″). A lid is unnecessary for these trays since feeder crickets can start hatching in a week or so. When it comes to grasshopper and katydid egg-laying trays, the keeper is usually looking at a deep container (substrate depth around 4.5″ or 12 cm) with a lid to maintain moisture over the three to six months of gestation. Of course, many of the non-feeder crickets take multiple months to hatch, eggs of some katydids can take more than twenty-four months (McMonigle 2005), and some grasshopper egg pods hatch in ten days. Katydids with flat or short ovipositors tend to lay inside plant tissue so an egg laying tray is useless. When laid in live plants it is best to keep the leaves or stems alive as long as possible.

The recommended mix for the egg-laying tray is sand with a small portion of peat. Sand allows for drainage and gaps for air, while peat retains moisture. These two are also easily accessible, relatively consistent, and inexpensive materials. Shredded coconut or potting soil substitute for peat well. Clay or composted materials should be avoided, as they include fine particles that can coat and suffocate easily if too damp.

Allowing soil to dry between watering is not necessary and will kill most eggs in whole or in part. Eggs and surrounding soil kept too dry or too wet, even for a short period, will doom the next generation. Too much water is almost always deadly; if the substrate is waterlogged every egg suffocates. Most eggs are dead if they dry enough to collapse. Once an ovum is indented more than a trace, adding water will not revive it. Substrate may protect ova from drying even if the substrate seems dry. Exposed eggs are dead before the keeper notices them, even if they are not chewed on. Both terrestrial species that dig and burrowing species often dig up the eggs accidentally in the confines of an artificial habitat. A layer of screen that

Though adults of the two species are similar in size, *Pediodectes haldemanii* (left) produces larger ova than does *Pediodectes nigromarginatus* (right).

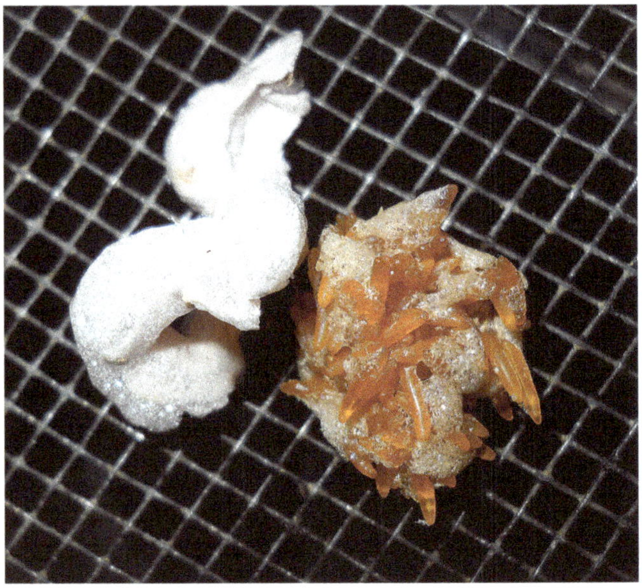

Ootheca failure in *Schistocerca nitens* looks similar to a mantis ootheca failure with a pile of eggs followed by a long tube of foam. The cause for this particular failure was an inability to find the egg-laying container. Moved immediately to a damp container, but every ova was dead.

SELECT SPECIES

Grasshoppers

The Caelifera include the familiar, far-jumping, day-active, often-flying, 'short-horned' grasshoppers that live in fields and deposit oothecae known as egg pods deep in the soil, but there are also burrowing species and phasmid-like arboreal species.

Aularches miliaris (Linnaeus, 1758)

This easily-maintained grasshopper, originally imported from Malaysia and Thailand, has been common in captive collections since the 1990s. There is only one species in the genus and its natural range includes much of China and India. The body colors look unreal—the adults especially look like they were manufactured by human hands from plastic and paint (and possibly explains why so many plastic grasshoppers are made in China). They produce a rasping noise and toxic, smelly foam when handled roughly. Full-grown specimens are not very big for grasshoppers, but are gigantic by insect standards. Females reach 2 3/4″ (7 cm) at maturity, while males are smaller, usually a little over 2″ (5-6 cm).

Life Cycle: A female produces three to five oothecae which usually contain 30-40 eggs each. They should be kept in damp, sandy soil with moisture level monitored regularly since they dry out easily. Eggs hatch after four to five months at 77° F (25° C). Nymphs grow to maturity after about four months and adults live another five months. For the first few captive generations, at least, the hatchlings mostly emerged around November and December (Larsen 2002).

Husbandry: This pyrgomorph does very well on a singular diet of blackberry and raspberry leaves (*Rubus* spp., i.e., brambles). A 5-gallon (19 L) glass terrarium with a screen lid works well and houses up to a dozen adult specimens. *Aularches* are a little more sensitive to drying out than

Adult *Aularches miliaris* in 2001

Dorsal view of adult *Aularches miliaris* © Tom Larsen

Adult *Aularches miliaris* © Tom Larsen

When disturbed, *Aularches miliaris* produce an odorous foam near the base of the legs. © Tom Larsen

Early instar *Aularches miliaris* © Tom Larsen

South African Pyrgomorphidae: *Dictyophorus spumans* (CC-BY-SA Alexey Yakovlev)

South African Pyrgomorphidae: *Maura rubroornata* (CC-BY-SA Alexey Yakovlev)

South African Pyrgomorphidae: *Phymateus viridipes* (CC-BY-SA Alexey Yakovlev)

South African Pyrgomorphidae: *Phymateus morbillosus* (CC-BY-SA Alexey Yakovlev)

field species, so a few inches of substrate helps maintain moisture in the enclosure. If the substrate is around three inches (7.5 cm) deep overall, the tank can also be used for gestating oothecae. Nymphs are highly resistant to overcrowding in terraria and are gregarious in the wild.

Newly hatched *Aularches miliaris* © Tom Larsen

Early instar *Aularches miliaris,* Feb. 2007 © Tom Larsen

Plains Lubber Grasshopper
Brachystola magna (Girard, 1854)
This grasshopper is found widely across the United States east of the Rocky Mountains, from North Dakota south to Texas and Mexico, with isolated populations as far as Minnesota, Iowa, Kansas, and Oklahoma (Froeshchner 1954). The plains lubber is the largest grasshopper in many of the states it resides in. It is always flightless and has tiny wings at maturity. There are a few different color forms often restricted to collection location, including a pink form, a green form, and a brown form. The blue and pink form pictured here is also common. Females and males are very different in shape, though both are large, over 2″ (5.5 to 6 cm). Males are thin with oversized back legs that give them the appearance of body-builders. Males also have a forked, pointed tip on the abdomen that resembles an ovipositor at a quick glance but there are no valves. Most other grasshopper males have a round tipped abdomen resembling a jockey cup. Females are wide-bodied and massive.

Life Cycle: It is difficult to get females to produce healthy oothecae in captivity. They usually drop them on the cage bottom and immediately chew on them if any are laid at all. A substrate of a rocky soil may be needed as an impetus, or the tendency to employ easier food items (dog food over romaine) may interrupt ova development. The oothecae or egg pods contain twenty eggs each and normally hatch around June (Capinera and Sechrist 1982). Nymphs mature in three months and adults live a few months.

Husbandry: A standard grasshopper enclosure with good ventilation is needed

Brachystola magna male and female specimens from B. Tomberlin, Hatari Invertebrates, 2005

Adult female *Brachystola magna* from near Roswell, New Mexico

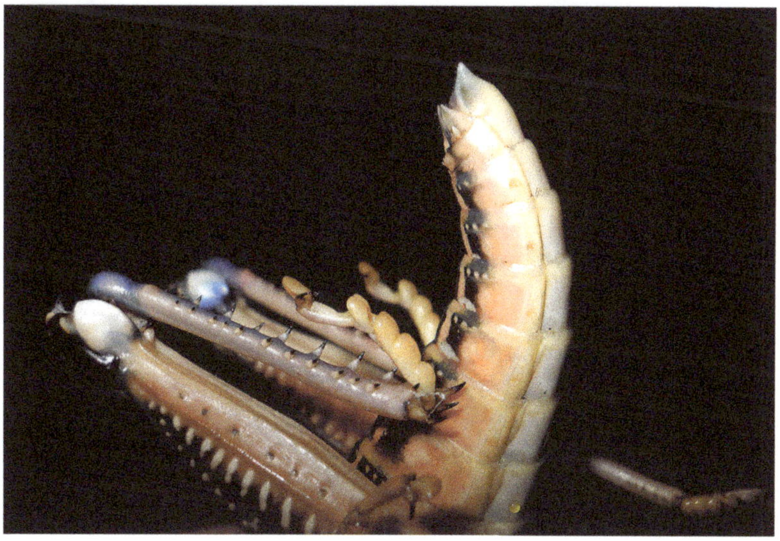

The male *Brachystola magna* abdomen has an unusual pointed tip used to fight other males.

Adult pair of *Dactylotum bicolor*

to keep specimens healthy. Soggy soil leads to dead specimens, so a bare bottom tank is suggested for small (less than 20 x 20″ or 51 x 51 cm) or deep enclosures, if glass-walled containment is used. Many different types of common produce from the grocery store may be eaten, while romaine lettuce is a good primary diet. Leaves of leatherleaf viburnum, cattails, and dry dog food are readily eaten by large nymphs and adults. The natural food is forbs (sunflower, cotton, ragweed, and lettuce) and dead or incapacitated insects. They are not predators but if kept with other grasshopper species their cagemates will disappear over time. Plains lubbers eat almost anything. This is a feature that makes grasshoppers difficult pests and sometimes difficult captives. They eat almost anything if their preferred food is not available, even if what they eat is no good for them, eventually kills them, or prevents them from developing correctly. One of the primary control strategies for swarms of various locusts is poison bait.

Rainbow Grasshopper

Dactylotum bicolor (Charpentier, 1843)

This small, flightless grasshopper is among the more familiar New World species because of its bright coloration. Males reach less than an 1″ (2.1 cm) at maturity and females are a little less than 1 1/2″ (~3.2 cm).

They look similar except for size and the tip of the abdomen. Both sexes have tiny, little wing pads at maturity.

Life Cycle: Up to two dozen eggs are deposited per egg pod in the fall. Nymphs wiggle to the surface in spring and go through five instars like most grasshoppers.

Husbandry: Oothecae are deposited in fine, sandy clay soil or creek loam (Capinera and Sechrist 1982). Common landscape plants fed on in captivity include firethorn (*Pyracantha*), rose, and raspberry. They also feed on various flowers including any hibiscus, but not the leaves. Despite some polyphagy in larger specimens, this species may not be possible to rear through a generation without *Baccharis* spp. foodplants. Captive success is very rarely reported. Like many grasshoppers, they die rapidly if kept too humid, but rainbow grasshoppers are more sensitive than far larger species. (Usually, tiny specimens are more resistant to excess humidity and limited ventilation.)

Migratory Locust

Locusta migratoria (Linnaeus, 1758)

The most familiar plague animal on earth has long been a common tenant of laboratories where studies of genes, motor activity, accessory glands, apolipophorin, peptides, trypsins, and research often unrelated to control occur. This and *Schistocerca gregaria* have long been reared in Europe as feeders for reptiles. Females, on average, weigh 50% more than males even at third instar, so they can be sexed by weight when reared under similar conditions. Females also undergo an unusual sixth instar (R. and F. Key 1936). The solitary phase of the locust has a green body with tan wings in the wild. Specimens kept under laboratory conditions nearly always develop into the gregarious phase with a brown body and longer, dark-splotched wings.

Life Cycle: Eggs diapause when females are kept under cool temperatures (~75° F / 24° C) and short day lengths. Gestation can take four or five months instead of two weeks. Growth of nymphs is less than a month at 86° F (30° C) for 16 hours a day. Females can wait long periods to produce egg pods if day length is long and temperatures are below 77° F (25° C). Adults live two to three months.

Husbandry: The migratory locust and *S. gregaria* are dry-loving grasshoppers that should be kept in full-screen cages. If kept in large numbers for feeders, they should also have a screen bottom to allow frass to fall out to a tray or newspaper or the moisture from the frass can overwhelm them. They eat various leaves and grains including romaine lettuce, rose leaves or hips, blackberry, various grasses, rolled oats, wheat, and sunflower seeds. For best production, temperatures should be 90-104° F (32-40° C) during the day and 77-86° F (25-30° C) at night. Under most circumstances, misting and drinking water should not be provided to *S. gregaria* nor *L. migratoria* (Bruins 1999).

Red-Legged Grasshopper

Melanoplus femurrubrum (De Geer, 1773)

There are about three dozen *Melanoplus* species in North America and *M. femurrubrum* is possibly the most common and widespread. It is forgiving when reared in the laboratory, so it is a good starter for those learning to maintain grasshoppers. It is also known as the garden grasshopper

Red-legged grasshopper, *Melanoplus femurrubrum,* adult female from Ohio

Arid lands spur-throat grasshopper, *Melanoplus aridus,* Arizona (CC-BY Katja Schulz)

Differential grasshopper, *Melanoplus differentialis,* is a large, common, urban grasshopper. This 4.2 cm female is from the Cedar Point (Ohio) artificial dunes.

Migratory grasshopper, *Melanoplus sanguinipes,* British Columbia (CC-BY Judy Gallagher)

Melanoplus femurrubrum wild-caught nymphs are often neon green.

This is normal coloration for a captive-reared *Melanoplus femurrubrum* nymph.

because specimens are a common sight in suburban gardens and border landscaping. The average North American from Canada to Mexico might find specimens in the backyard. Nymphs are bright green and found solitary in the field, even when small. In enclosures, nymphs are more often brown and do not cannibalize or damage each other in close quarters. Adults are gray and tan with red tibiae (rather than red femora).

Life Cycle: Eggs are laid by females from mid-summer to frost, after which the females die. In captivity they can live a little longer but often die before winter. Males are significantly smaller than females and expire months earlier. Eggs hatch around April the following year. There are five instars, taking about three months at 75° F (24° C), and probably much faster in hot, sunny fields.

Husbandry: Red-legged grasshoppers are easily kept like the large *Romalea* on romaine lettuce from the grocery store when small, moving to bramble, rose, or leatherleaf viburnum when larger. (There are many other *Viburnum* species that grasshoppers will not eat.) They mostly eat grass in the wild, but that can be a little hard to find in good shape in early spring when they tend to hatch in captivity. Gestation at 72-75° F (22-24° C) is suggested because cold diapause is more difficult to monitor and higher temperatures could cause an even earlier hatch and less food availability. Leatherleaf viburnum leaves are accepted and are easier to keep fresh than grass or lettuce. A tall deli-cup (32 oz., or 1 liter) filled with damp sand, mixed with some potting soil or peat, works well for egg laying. Females usually have no problem finding the container if a stick is placed at an angle from the bottom of the enclosure to the top and resting against the lip of the container. After a few weeks, when a few test holes are observed (pits are signs of egg laying), a lid is placed on top and the container is moved to a shelf. If females are still healthy a new container for laying should be placed. Oothecae are loose and do not seem to hatch well if dug up. However, grasshopper oothecae are not very solid like mantis oothecae and breaking them apart does not necessarily kill the

This adult male *Melanoplus femurrubrum* was damaged by a lawnmower as a nymph and the eyes were smashed in, but they recovered, Medina Ohio.

affected ova. In nature and captivity the tiny males mature when the females are two molts behind. Only the early bird gets to sire offspring. Female *M. femurrubrum* can be large with a 3 cm body length, but the hatchlings are surprisingly tiny. Also, despite the diminutive size of individual eggs there are only about two dozen per egg pod.

Horsehead Grasshopper

Pseudoproscopia scabra (Klug, 1820)

Proscopiidae is a peculiar family of grasshoppers found only in Central America and northern South America. Members of this group bear a striking resemblance to stick-like phasmids, but the unique face and very large back legs give them away. As in other grasshoppers, the rear legs are adapted for jumping. Specimens jump, but it is not a primary defensive reaction. They usually play dead. Older nymphs and adult males can jump twelve inches horizontally. Mature females lack the ability.

Pseudoproscopia scabra is found in Brazil, French Guiana, Peru, and Venezuela, and has been in captivity for decades. Adult males are around three inches long (~7.5 cm). They are colorfully striped in bright aqua-blue and dark gray. Female *P. scabra* are large, brown animals averaging six inches (15 cm) in length. A more recently maintained species, *Pseudoproscopia latirostris* (Brunner von Wattenwyl, 1890) from Huanuco, Peru, seems to be a good candidate for long-term care. Females look similar but are smaller (10-11 cm). The 7 cm males look quite different, only because they are pea green in color. It is overall very similar to *P. scabra* in life cycle and care but the egg pods contain only 20 to 30 eggs (Hes 2020). The various species and genera of proscopiids look very similar with differences mostly in the size of the female and color of the male. I first observed specimens of *Stiphra* sp. at *World of the Insect* in 2010. Stock was collected in 2000 and is currently (2021) displayed at various American insect zoos. Males are marked with green like the male *P. latirostris*. Females are large, mottled brown, and remarkably similar in appearance to *Pseudoproscopia* females.

Life Cycle: Eggs take about four months to hatch and then females need half a year to mature. The miniature males take about five months to mature. Growth takes almost as long, even though they are maybe a quarter the mass. Development time disparity is not a problem since both genders are long-lived, often surviving close to a year.

Husbandry: A glass terrarium of 10-30 gallons (38-114 L) with 4″ (10 cm) of damp sand mixed with dirt and a screen

The unusual face of proscopids give rise to the name 'horse-headed grasshoppers.' This is a female *Pseudoproscopia latirostris*.
© Ondřej Hes

Adult pair of *Pseudoproscopia scabra*

The small, colorful male of *Pseudoproscopia scabra*, 2002

Adult pair of *Pseudoproscopia latirostris* (female's massive abdomen is out of the shot) © Ondřej Hes

Pseudoproscopia latirostris buries its very long abdomen while laying eggs. © Ondřej Hes

Stiphra sp., face of the female
World of the Insect, 2021

Coloration of the adult male *Stiphra* sp. changes from mottled brown on the final molt.

lid serves well as an enclosure. Ventilation is a necessity. A metal screen lid is necessary if a basking heat source is applied from above. A full screen cage is not recommended if ambient humidity is less than 40%. Half a dozen specimens for every 10 gallons (38 L) of space can be maintained safely. The enclosure should be a foot (30 cm) high or better for the large females to undergo late instar molts. They are not aggressive eaters so they normally will not nibble on or damage their neighbors, but they can accidentally dislodge a molting animal. Also, too many specimens together can prevent some animals from reaching food and make it difficult for the keeper to maintain a consistent supply of food. Optimal temperatures are usually around 72° F (22° C) at night to 81° F (27° C) during the day, but temperatures above 75° F (24° C) and day-night cycles are not required for success.

A container of substrate for oviposition will be necessary if a screen cage is used, but females can have a difficult time finding a container of substrate. Eggs laid on the floor of the cage dry out within minutes. Even if encountered and buried immediately by the keeper, the survival rate of eggs laid in the open is minimal. The female drives her entire abdomen deep into the soil to lay an ootheca consisting of thirty to fifty eggs. The abdomen does not telescope during oviposition like most grasshoppers. However, her abdomen is exceptionally long which allows her to bury the egg pod about three inches (7.5 cm) deep. Ova require three to four months to hatch, and the substrate must be kept damp the entire time. Conversely, excessive dampness can suffocate the ova because the substrate will go anaerobic. Nymphs begin the first molt within minutes of reaching the surface after hatching.

Pseudoproscopia scabra are not omnivorous like most grasshoppers and after hatching are kept more like a phasmid than a grasshopper. They feed exclusively on fresh leaves of select plants including the usual phasmid staples: rose, blackberry, and raspberry. Specimens can be difficult to discern from twigs, so foodplants with fuzzy or green stems can help to prevent animals from being discarded. Branches of the foodplant should be placed in water to keep leaves fresh. The water container should have a lid with a punched hole or paper towel to prevent small animals from climbing inside and drowning. Moreover, the container dimensions should be large enough to keep the cutting supplied for days and should be wide and heavy enough not to fall over. Floral spikes and small vials hold too little water and are difficult to keep in place.

Eastern Lubber Grasshopper
Romalea microptera (Beauvois, 1817)

The word 'lubber' refers to a heavy, clumsy, or awkward fellow. All are good words to describe the eastern lubber grasshopper, one of the heaviest of the United States' insects and one of the heavyweights of the grasshopper world. A large female measured from the head to the tip of the abdomen can reach more than three inches (8 cm). The wings are shorter than the body and add no length. Lubber grasshoppers are unable to fly, are slow moving, and have limited jumping ability. The clumsy and awkward demeanor makes them an endearing and easily handled pet. (Many other large grasshoppers tend to have advanced fight-and-flight skills which makes handling uncomfortable for the keeper or dangerous to the animal.) Though probably not the original reason for the name, the word 'lubber' also means 'a clown' and is descriptive of the flamboyantly colored outfit of *R. microptera*. Spots and stripes of red, yellow, black, and orange paint the armored suit. This wonderful little clown has entertained children at insect zoos around the world since the 1970s.

The eastern lubber is usually listed as the namesake of its own family, the Romaleidae, which includes several primarily New World and often large and colorful species that have two immovable spines at the end of the tibia. The older taxonomic scheme lists the Romaleinae as a subfamily of the Acrididae. This is the only species in the genus *Romalea* although some texts mention *'R. guttata,'* which is a synonym. The closest relatives are

Romalea microptera,
black form from Florida
© J. Byrne

species in the genus *Taeniopoda,* which are differentiated because the males can fly (though seemingly only in certain wild situations, since they do not fly in captivity). Sometimes *Taeniopoda* is listed as a subgenus of *Romalea*.

The species name *microptera* means 'tiny wing.' The wings are short compared to flying grasshoppers, though *R. microptera's* wings are huge compared to the tiny round wings of the plains lubber. *Romalea's* hindwings are dark pink and black and can be raised up in a warning or deimatic display. Forewings normally stretch halfway down the abdomen. The length of the tegmina can be longer in some populations and varies according to how full the abdomen is (when wing length is expressed in relation to the abdomen rather than the pronotum).

This species is found across a huge area of the United States, from North Carolina south to the tip of Florida and west through Texas. As is common for a flightless invertebrate with a wide range, there are several geographic size differences and color variations. The two main color forms are usually referred to as 'black' and 'olive.' The black form has yellow edging which resembles nymph coloration and looks like the related, southwestern *Taeniopoda eques*, but without the yellow vein-like markings on the rear femur and tegmina. *Romalea microptera* color forms can interbreed in captivity, but otherwise do not normally throw nymphs of alternating color forms. Nymphs of both forms are primarily black, but the black form nymph's margins darken to red after the first molt. The olive morph nymph retains yellow edging through subadult. There are many different geographic colors

Adult pair *Romalea microptera*

Pink form *Romalea microptera*, February 2006

Romalea microptera subadult

Romalea hatchlings, late June 2006

including an unattractive intermediate of black and olive and an attractive pastel orange variation of the olive form.

Food: *Romalea's* large size is misleading because it eats less than many grasshoppers a third its size. It will nibble on a variety of leaves though very few can be used as a primary food. Some common plants including privet, rose, and blackberry are eventually eaten, but romaine lettuce is the only common food eaten with gusto. They seem to have almost no appetite when fed leaves other than romaine, but less common or seasonal foods like daffodil, dandelion, purple dead-nettle, and cattail reeds can work. Wheat or millet seeds can be planted in a shallow tray and used as food when the sprouts reach an inch or two. Dry dog food, cat food, and carrots can be added to provide additional nutrients and to stretch out the primary food supply. Romaine lettuce can be used as the only food if preferred. Nymphs and adults have the same general diet but fresh greens should be available daily for the first instar to ensure optimum survival. Only subadults and adults eat most types of dry dog food pellets, because they have more interest in protein or are less picky. Small nymphs prefer to feed on vegetation but they go after the green 'vegetable' pellets in Beneful brand dog food. (Palatability is related to vegetable content or color, not sturdiness of the pellet, because other, far softer pellets in the Beneful are refused.)

Behavior sets *Romalea* apart from the closely related horse lubber because it lacks a tendency towards cannibalism. Even when starved and overcrowded, it rarely bothers tank mates. Horse lubbers walk over and eat each other from behind, a rather ghastly behavior since the victim appears quite unaware and walks around with half its body missing, even if it is just a head and two front legs. *Romalea microptera* also tends to grow larger than *T. eques* in captivity.

Romalea microptera can be considered a nuisance pest in gardens. Gardeners (those who lack love for beautiful invertebrates) tend to be frightened upon seeing a grasshopper the size of a small bird, even if there is hardly any damage. Eastern lubbers are very slow-moving, harmless, occur in low numbers, and have large body mass, so control advice is to collect them and use a 'smashing device.' They are supposed to cause some damage to citrus and daylilies (plants originally from tropical Asia). In the lab they were offered orange leaves, which were not taken readily but were half eaten after a few days in the cage. Daylily leaves were also offered but only nibbled on. Daylily flowers were half eaten in a few hours but never finished.

Life cycle: Egg cases are laid two to three inches deep in the soil. The female buries her telescopic abdomen up to the back legs and forms a compact, foam-capped ootheca underground. Though glued together, a minimal amount of force shatters it into separate eggs. (It does not compare to the structured oothecae of mantids or roaches.) *Romalea* females make 3-5 egg cases and each contains 25-40 eggs. Grasshoppers hatch after four to five months at room temperature, 72° F (22° C). Hatchlings need some moisture and should be fed fresh greens every day or two. Nymphs molt every twelve to twenty days until the final molt, which can take more than a month. Adult females feed for four to eight weeks after the final molt before laying the first egg pod.

The following dates represent a single lab rearing from 2006 and are provided as a timeline reference. Ova gestation temperature averaged 72° F (22° C). Nymphs and adults were kept around 78° F (26° C).

February 14	Eggs Laid
June 21	Hatching
July 4	First Molt
July 15	Second Molt
July 27	Third Molt
August 12	Fourth Molt

Husbandry: The most difficult part of the eastern lubber life cycle is gestation of the ova. Exposed, dry conditions wipe out every egg in as little as a few hours and hatching rates can be extremely low without a repeatable and dependable methodology. The best method requires no upkeep—using a clear 32 oz. deli-cup and lid with no air-holes. The container is filled 2/3 with inexpensive play sand mixed with 10% coconut fiber or peat. It should be very damp but not waterlogged. The surface should always stay damp enough to clump and not separate when pinched or it is too dry. After a few oothecae are laid (this can be difficult to observe) any frass is removed from the surface and the lid is placed on top with a date label. The surface moisture can be checked from time to time but there are too many pinholes in the lid if it must be moistened more than once in four months. Hatchlings must be moved to a vented cage with food within a day or two or they will die from the humidity and lack of ventilation.

A 10-gallon to 30-gallon (38-114 L) terrarium with glass sides and a metal screen lid is suggested for rearing nymphs to maturity. As they grow to maturity specimens should be thinned out to about six for every 10 gallons (38 L) of space. Foodplants should be readily available but not so numerous as to block the view. It is important to have angled surfaces such as branches, so nymphs can molt successfully when kept in a smooth-walled container like an aquarium or plastic small animal cage. They will hang upside-down from the branch or lid to molt. Virtually defect-free molts can be achieved if they are kept in a screen-walled cage such as those used for chameleons.

Good ventilation is incredibly important as stagnant, damp air will lead to death in days to weeks. The lid, or at least one side, should be made of window screen. Unless the food or atmosphere is extremely dry, daily misting is unnecessary.

American Bird Grasshopper

Schistocerca americana (Drury, 1770)

The bird grasshoppers in the genus *Schistocerca* include some of the most impressive acridids in the United States. They are excellent fliers, easily flying more than 100 yards when disturbed (Ball et al. 1942). Flying off into a tree is usually a good defense, though I have seen *S. obscura* alight straight into golden orb weaver webs, high off the ground, between tree trunks. (*Nephila* probably have the only webs capable of catching these grasshoppers.) In an enclosure, the fright and flight behavior can be very loud and jumping specimens can move a small cage. The various species reach a maximum size between 2″ and 3″ (5-7.5 cm) in length, though each species can demonstrate variability in size. The infamous swarming desert locust, *Schistocerca gregaria*, is the only Old World member of the genus and is commonly bred as reptile food in Europe. Recently, some keepers have begun to breed North American bird grasshoppers for this same purpose (Byrne pers. comm.).

Life Cycle: In nature, there are two generations per year. The adults are long-lived and overwinter beneath leaf litter and logs. They become active again in spring, producing the next generation in summer. In captivity, as in the southern U.S. where generations overlap, *Schistocerca* can be kept active year-round. Egg pods with about fifty eggs are laid in the soil and hatch in three to four weeks. Like most grasshoppers, the young pass through five instars before reaching adulthood.

Husbandry: Bird grasshoppers thrive in a full screen enclosure, but can also be kept in tall glass terraria with adequate ventilation. Multiple high branches and

A tiny adult female *Schistocerca nitens* (4 cm) at World of the Insect, 2021

Female *Schistocerca nitens* (8.5 cm head to wing tips)

Schistocerca obscura, Georgia 1996

screen surfaces are important for large nymphs to molt safely as they hang upside-down. They cannot chew through metal screen but may eat holes in nylon screen or muslin cloth. Damage to the wing tips is usually from metal screen and flight attempts, not wing biting. These sun-loving grasshoppers do best with a heat lamp above the enclosure, which increases their growth rate, feeding, and activity level. They also bask under LED lighting but the lack of heat may make this pointless. *Schistocerca* will eat a wide variety of plants, from tree leaves to grasses, and devours romaine lettuce. Other reliable staples include oats, wheat germ, and bran. Larger specimens will eat pelleted foods for guinea pigs or birds.

Adult *Schistocerca* may spend a few weeks to months after the last molt before mating, depending on the species, but will mate frequently after they have eaten enough. Females deposit egg pods in containers of sandy soil. Once signs of laying are noted, the laying containers can be placed inside a large container tub with limited ventilation, or covered individually, to retain moisture. When the young start to hatch, the container can be placed back inside a rearing enclosure.

Horse Lubber Grasshopper
Taeniopoda eques (Burmeister, 1838)

The species name *eques* comes from the Latin word for 'horseman' or 'knight' and gives rise to the common name. It is difficult to say if the species name refers to the massive size, the horse-like face, or some other association with horses. It ranges from Arizona to Texas in the United States and is found south all the way to Costa Rica. There are eleven similar *Taeniopoda* species found from Mexico to South America, including *T. auricornis, T. burmeisteri, T. centurio, T. picticornis,* and *T. superba*. Considering the minimal differences, it is not surprising many of the *Taeniopoda* were originally described as *Romalea*. *Romalea microptera* and *Taeniopoda eques* can hybridize in the laboratory to produce fertile offspring. *Taeniopoda eques* is a member of the Tribe Romaleini in the lubber grasshopper Family Romaleidae. The Romaleidae is small in number but contains many of the largest grasshoppers, including the largest of all: the birdwings in the genus *Tropidacris*.

Horse lubbers are gaudily colored and one of the few insects to employ every one of the usual warning colors (black, red, white, yellow, and orange). Nymphs are black with yellow and white margins. Adults are black with orange margins. Contrasting red and black underwings are flashed in warning if large predators (such as humans) approach. Wing flashing is common in the wild but seldom practiced in captivity. Birds, mammals, and even toads refuse to eat them. Predatory invertebrates like tarantulas, centipedes, and mantids feed readily on lubbers—they apparently do not mind the taste and probably cannot see colors anyway.

Life Cycle and Husbandry: Females deposit egg pods containing around forty eggs, 3-5″ (8-13 cm) deep in damp, sandy soil. They retain eggs while searching for a good area, but eventually drop the eggs in a lump on any surface. It is important to provide egg-laying substrate before this occurs because exposed eggs desiccate rapidly and other grasshoppers in the cage will chew on or devour exposed eggs.

Taeniopoda reticulata, the purple lubber in Cahuita, Costa Rica © Ondřej Hes

Taeniopoda eques, late instar nymph, 2007

Some foam is formed around the eggs, but primarily as a cap, so it is a loose mass and only well protected below the surface. If the eggs are never allowed to dry out (deep substrate may become dry without drying eggs) they hatch after six to twelve months at room temperature (72° F or 22° C). For ova survival, sandy substrate should always be kept moist but not clogged or saturated with water.

Taeniopoda eques emerge two to six at a time for a week or two, not as a single group like *R. microptera*. They slowly worm their way to the surface but eventually die if they cannot break free. The eggs should not be reoriented since the hatchlings tend to go straight away from the top of the egg and can be doomed if the egg is facing down. Avoid digging up eggs. Those that hatch against glass or clear plastic often do a few loops resulting from a loss of direction due to the light but tend to make it to the surface anyway.

The molt begins when they are no longer under compression. They do not molt against the glass under substrate (proving light is not the factor that instigates the molt), but will molt in the dark if they end up in an underground cavern in the sand. Prior to hatching, make sure to tamp down sand to prevent cavities large enough to trap emerging young. The molt begins within minutes of breaking free. Compression will postpone molting for up to a few hours, but they will die in the old exoskeleton if they do not make it in time.

The wormy hatchlings have full-length, curvy legs but cannot walk until they undergo the first molt. While eggs can be kept in a permanently damp, barely ventilated container, nymphs must be moved to a screen or part-screen cage within hours to survive.

Baby lubbers feed on leatherleaf viburnum, wandering Jew, raspberry, rose, etc., but require a supply of fresh romaine lettuce to do well. Old lettuce should be discarded and replaced daily. The survival rate is high only with a constant supply of fresh romaine. Any type of dry cat or

Taeniopoda eques oviposition, Sept. 2007

Taeniopoda eques hatchlings in May 2007

Female *Taeniopoda eques* with an inconsequential dent in the head, Aug. 2007

Taeniopoda eques male with short wings

Taeniopoda eques male with long wings

dog food pellet is also eaten but cannot be used as the main diet.

The post-hatching molt is the only one that can be successfully completed on a surface. Nymphs can molt successfully the subsequent one or two molts as they cling from a smooth surface. During later molts the nymphs must hang from a stick or screen side—they will become deformed or stuck if they fall or molt on the bottom of the cage. If an aquarium is chosen over a screen cage, place it on the end or side so the screen lid becomes a wall. A branch stretching from the bottom corner to the top corner is helpful. Each molt requires two to three weeks of feeding and there are about half a dozen molts in total. Molting surfaces become more and more important and molts grow more dangerous as they get closer to maturity, because the animals are heavier.

Horse lubbers develop wings on the molt to adulthood. Adults are thick and sturdy and can exceed two and half inches in length (~7 cm). The flightless females are bulkier, and the abdomen sticks out past the wings. Males can have short wings like a female, but more often have wings that exceed the length of the abdomen. Males with long wings are capable of short flights. Both sexes live three to four months after reaching adulthood.

This species is extremely cannibalistic and famous for its tendency to accumulate on highways. They are attracted to their own dead as a food source and in turn are run over while feeding, so the attractive meal continually adds to the pile of death. If they are hungry, adults just walk over and eat each other, starting from the rear and often attacking in a ravenous but terribly slow-moving group. This leaves the victim only a sad little head and front legs to walk around. Every day or so a new victim is singled out and consumed by the group. Fortunately, it is not necessary to keep them in individual enclosures. If fresh romaine lettuce is regularly included in the diet the problem of cannibalism is very rare. Of course, for a keeper using the wrong food, the resulting group behavior can be rather disconcerting.

Comparison of sibling male *Taeniopoda eques* with different wings

Adult *Tropidacris collaris,* 2003
© Tom Larsen

Tropidacris collaris nymph, 2004
© Tom Larsen

The flying wings on *Tropidacris collaris* are somewhat blue.
© Tom Larsen

BLUE-WINGED GRASSHOPPER

Tropidacris collaris (Stoll, 1813)

This is arguably the largest grasshopper in the world, though there are two other species in the genus, *T. cristata* (red-winged grasshopper) and *T. descampsi,* that fall in a similar size range. The wingspan can be over 8″ (>20 cm)—supposedly one old museum specimen has a bullet hole in the thorax from being shot down by an ornithologist hunting birds. It is a member of the Family Romaleidae like the North American lubbers. *Tropidacris collaris* has been reared as a pet in Europe since 1990, when specimens were brought back from Guyana (Rogez 1996). It remains one of the most popular species in European collections today. It is native to most of South America, Central America, and into Mexico. Nymphs are gregarious when young, roaming in smaller groups as they grow, becoming solitary as adults.

Life Cycle: The length of time from oviposition to death from old age can easily last three years in captivity, which is rather long for a grasshopper. Eggs take surprisingly long to hatch, eight months to a year, but the hatch rate is usually excellent. Maturity is reached in six months, which is a little slower than the average grasshopper. The adult stage lasts ten to eighteen months.

Husbandry: They eat many other plant leaves in addition to the blackberry leaves they are commonly reared on. Their natural foods include plants from the Family Melastomataceae (Rogez 1996). Other leaves they are known to feed on outdoors include coconut, banana, citrus, coffee, tapioca, corn, sorghum, and sugar cane (Larsen 2020). They require excellent ventilation, so a screen cage and regular misting is suggested. The terrarium must be tall and large to prevent possible complications during molting that are commonly seen in small enclosures. The suggested temperature during the day is 77-86° F (25-30° C). High humidity and temperature can be maintained in a screen enclosure by placing a small water heater in a screen covered water container. Substrate for egg laying is usually a peat and sand mixture about 4″ (10 cm) in depth. The long incubation of oothecae requires some investment in keeping the substrate damp but not wet. Eggs should not be moved (Krejci 1999).

STINK LOCUST

Zonocerus variegatus

This African pyrgomorph, like *Aularches*, is brightly colored and can release a very smelly secretion when disturbed. Idowu (1997) found praying mantids were not affected, but ants and lizard were repelled by the secretion. Experimental results suggested the grasshopper invests in the repulsiveness of the secretion more than the toxicity. Nevertheless, pyrgomorphs sequester chemicals from their food plants so specimens on captive diets often smell very little if at all. In 2005, on TV3's reality show *The Farm,* some hungry contestants tried eating them and were sick for days (Larsen 2011b).

Males reach ~1.5″ (3.5-4 cm) and females mature at about 2″ (4.5-5 cm). The other member of the genus, *Zonocerus elegans,* is larger (females up to 7 cm) and usually possesses tiny wings. *Zonocerus variegatus* usually have full-length wings—but some populations are brachypterous (short-winged). The short-winged *Z. variegatus* has longer wings than the short-winged *Z. elegans*, but wing length is useless for identification. One consistent difference is that *Z. elegans* has orange bands up and down the antennae, whereas *Z. variegatus* has two or three such bands near the apex.

Life Cycle: Egg pods hatch after three to four months, but eggs laid at different times by the same group of adults in terraria usually hatch in concert. Nymphs mature after three-and-a-half to four months, with males maturing a few weeks earlier than females—then adults live another four to five months.

Husbandry: Preferred temperatures are 82-86° F (28-30° C) during the day to as low as 68 F (20° C). This is not an easy species for the beginner because the eggs need a very consistent humidity and temperature, or they will not hatch. In nature they live in fringe habitats, but are a common pest of corn and cassava (*Manihot utilissima*). Captive specimens can be fed beech, oak, rose, blackberry, and raspberry, but thornless varieties should be chosen since they are clumsy and easily damaged or even killed by thorns. They also feed on slices of banana, apple, grape, mango, and melon (Larsen 2011b).

Zonocerus variegatus female, 2005 © Tom Larsen

Zonocerus variegatus pair, 2005 © Tom Larsen

Wingless mountain grasshopper,
Booneacris glacialis
New Hampshire
(CC-BY Judy Gallagher)

Scudder's short-winged grasshopper
Melanoplus scudderi
Virginia
(CC-BY Judy Gallagher)

Southern marbled grasshopper
Spharagemon marmorata picta
Florida
(CC-BY Judy Gallagher)

Pale toothpick grasshopper
Paropomala pallida
Santa Catalina Mountains, Pima County, Arizona
(CC-BY Katja Schulz)

Longheaded toothpick grasshopper
Achurum carinatum
Florida
(CC-BY Judy Gallagher)

Glassy-winged toothpick grasshopper
Stenacris vitreipennis
Polk Co., Florida
(CC-BY-SA Mike Leffler)

Brazilian flag grasshopper, *Chromacris psittacus,* Costa Rica
(CC-BY-SA Pavel Kirillov)

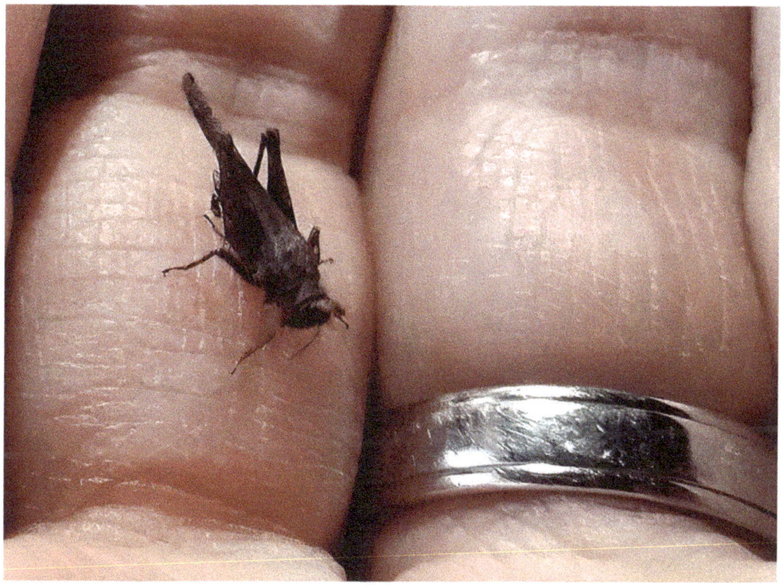

Pygmy grasshopper, *Tettigidea lateralis,* overwinters in Grand River, Ohio, as an adult.

Pygmy grasshopper
Australia
(CC-BY Jean and Fred Hort)

Xiphiola cyanoptera
Ecuador
(CC-BY-SA Pavel Kirillova)

Horse lubber
Taeniopoda eques.
Santa Catalina Mountains, Arizona

Eastern lubber, nymph
Romalea microptera
Florida

Tropidacris sp., nymph
Costa Rica

Taeniopoda reticulata, Costa Rica (CC-BY-SA Pavel Kirillov)

Painted grasshopper and nymph, *Poekilocerus pictus,* India (Haneesh K. M.)

Heide sp.
Australia
(CC-BY Jean and Fred Hort)

Horsehead grasshopper
Pseudoproscopia scabra
French Guiana
(CC-BY-SA Bernard Dupont)

Horsehead grasshopper
Ecuador
(CC-BY-SA Pavil Kirillov)

Acheta domesticus nymphs and adults can be kept in a single rearing cage. Only the adults are black in the hypermelanistic form.

Normal colored adult male *Acheta domesticus*

Crickets

True crickets include the familiar, omnivorous, night-time singers, but also arboreal predators, day-active, silent, and underground creatures.

HOUSE CRICKET

Acheta domesticus (Linnaeus, 1758)

This is the most widely kept orthopteran in captivity. It has been used as a common staple for feeding reptiles since the 1950s and in recent years is being reared for human consumption. It probably originated in tropical Africa since other *Acheta* species are found in northern Africa and the Arabian Peninsula. While it is slightly resistant to cold, it is tropical. Feeder shipments in mid-winter in cold areas often arrive dead and it can be difficult to acquire this live food during particularly bad cold snaps. It has long flying wings, but lacks the ability to fly.

As a pet *A. domesticus* is easy to acquire in almost any country. For me it sings one of the most melodious songs. While everyone has a different taste in cricket music, Stanek (1969) also mentions an appreciation for the house cricket's pleasant song.

One thing I observed from years of keeping this species is the ability to learn and unlearn inherited behaviors. After half a dozen generations, specimens would always jump off of the egg carton when it was picked up. Escape behavior is normal for wild crickets. Feeders bought at the store seem to cling to the carton for dear life, even if you walk across the room holding it. In just a few generations they can learn, unlearn, and relearn behaviors. In cricket factories, animals that jump away are not used to start new cultures, while in my cultures and in nature those that do not try to escape are the first to be consumed.

Life Cycle: The eggs hatch after 10-15 days and they grow to maturity in five to nine weeks with the faster growth occurring at temperatures from 85-90° F (29-32° C). Adults live only about a month or so, but a female can produce nearly a thousand eggs. Ova usually will not hatch below 75° F (24° C). Over time and many generations, the temperature parameters and longevity can be changed. I line bred a stock in cooler temperatures over fifteen years using a single enclosure—a 48 x 18 x 21″ glass terrarium with a screen lid. After the first ten years (~40 generations), the eggs hatched well at 72° F (22° C) and the adults would live three to four months.

Husbandry: Warmth and good ventilation are especially important for culturing the house or hearth cricket. Usually, a warm spot light is suggested for each enclosure. With incandescent bulbs going the way of the dinosaur, a ceramic heater or infrared bulb may be more practical than halogen for heating. Under-cage heat tape with a thermostat set to 80° F (27° C) is a secondary or dual option to reduce hatching time and increase yield. Although the ova require damp conditions, warm and dry is best for nymphs and adults; but they need a source of moisture to drink. Water can be poured onto a sponge-filled dish or, if there is substrate, to a small corner of the media. An open dish would result in multiple drownings and a gravel dish would result in a rotting corpse, food, and phorid maggot mix. If egg-laying media is kept in the cage and the substrate moisture drops, adults will dig up the juicy eggs and eat them. Cricket

Adult female *Gryllodes sigillatus*

Adult male *Gryllodes sigillatus* with common coloration

Adult male *Gryllodes sigillatus* with black tegmina

water jelly (polymer that holds water) or slices of carrots or potatoes can be used to provide moisture and reduce this problem. Egg trays kept in the main enclosure may also be protected with metal screen placed directly on the soil. Poultry mash or dry dog food are often used as a primary food from hatching pinhead to winged adult. Various companies make cricket food, but they function no better and tend to be expensive in tiny quantities. For serious cricket breeders, Purina® has offered 45 lb. (20 kilo) buckets of cricket chow for half a century (Dunn 1993). Cannibalism can take a great toll on specimen numbers if food or water is limited or delayed. Large numbers of molting animals can be consumed and all signs of their passing devoured. Crowding also increases cannibalism, but not as significantly, and is reduced by the paper egg cartons that add significantly to the usable surface area of the enclosure. Although drinking water is very important to have, water too much and they rot. House crickets are among the easiest orthopterans to breed, but neglect will greatly reduce their numbers.

Banded Cricket

Gryllodes sigillatus (Walker, 1869)

Also known as the tropical or Indian house cricket, this species replaced *A. domesticus* in most pet shops for a few years after the 2010 cricket plague, because it is immune to the cricket paralysis virus affecting *Acheta*. Males especially are tiny-bodied, which may be why use as a feeder is subsiding. One benefit of this species is that specimens are less likely to chew on flowers and leaves in planted terraria. The larger females have nymph-like wing pads, while the males have large round tegmina for singing, but no flying wings beneath. The male's chirp is not nearly as loud as the house cricket, but it is a long, pulsating trill that for me is less pleasant. Males usually have pale, light-tan wings but some have black wings. The banded cricket lives outdoors near human structures in tropical areas worldwide, including the warmest areas of the United States. It was probably restricted to Asia before human beings came along.

Life Cycle: Banded crickets supposedly require warmer-than-room-temperature for the eggs to hatch, but I have experienced no problems a little below 75° F (<24° C). Females rarely lay more than a hundred eggs. Males mature after two months and females after two and a half to three months at 75° F (<24° C).

Jamaican Field Cricket

Gryllus assimilis (Fabricius, 1775)

After the 2010 cricket crash this species became widespread as a feeder in Europe and the United States, but I never encountered it at northeast shops or reptile shows. Many pet bug keepers and even some reptile keepers blamed the deaths of their pets on this species. Still, this is not an entirely new story since house crickets opportunistically feed on sick, dying, or dead pets. The adults are big and pretty with orange markings. Jamaican field crickets naturally occur in warm areas of North, Central, and South America, and the West Indies. Specimens can be collected in Florida and Texas. A similar looking cricket in western state and Mexican populations is *G. multipulsator* (Weissman et al. 2009). *Gryllus assimilis* remains a quite common feeder cricket in Europe.

Gryllus assimilis, adult pair © Ondřej Hes

Gryllus bimaculatus: even as a large nymph it does not yet have the marked wings for which the species is named. © Ondřej Hes

Life Cycle: Eggs hatch in 11 days normally.

Black Field Cricket

Gryllus bimaculatus De Geer, 1773)

This species is larger (to ~1.5″ or 3.5 cm) and more aggressive than *A. domesticus*, so fewer animals are needed to keep large predators fed. It is also likely to eat wimpy predators and has a louder song than any other common feeder. It is dark brown to black, with trademark yellow markings at the base of each tegmina (*bimaculatus* means 'two markings'). This cricket cannot be distributed in the U.S. because it is not native, but it has long been a popular feeder in its native Europe. It is wide-ranging, native to India, China, and West Africa and has long been considered an agricultural pest (Stanek 1969).

Crazy Red Cricket

Gryllus locoroho Weissman & Gray, 2019

This has been commonly bred and used as a feeder in Europe since the late 1990s, but the species was only described in 2012. It was traded under the label *Gryllus assimilis* for the first twenty-five years, but seemed a little different. Since it was described as a new species, which is not the U.S. native *G. assimilis*, it is not supposed be shipped or used as a feeder in the U.S. Interestingly, the type specimen is from California, while the original collection locale is uncertain.

Ethiopian Cricket

Homoeogryllus xanthographus Guérin-Méneville, 1847

The original generation was imported from Ethiopia or Kenya in 2005, and was first erroneously traded as *H. indicus* by hobbyists. The body length is up to 0.8″ (2 cm) and relatively slender. Males have prominent wings and females have long ovipositors. Coloration is dark brown with tan to light brown wings. The singing of the males is very unusual because it resembles the song of a small bird, rather than an insect.

Husbandry: Ethiopian crickets do best in a relatively spacious tank, 20 x 12 x 12″

Homoeogryllus xanthographus, adult male (top) and an unusually dark female (bottom) © Ondřej Hes

(50 x 30 x 30 cm), with good ventilation and not too many specimens. Egg cartons work well for shelter and climbing and should be removed when dirty. A shallow layer of peat can be used as substrate. A deeper tray of sand mixed with peat can be provided for oviposition. The laying tray should be kept damp at all times and should be at least 2″ (5 cm) deep. Like common crickets they are normally reared on fish flakes, vegetables, and dog food. The temperature should be 75-82° F (24-28° C). Higher temperatures should be avoided while slightly lower temperatures are tolerated for some time.

Life cycle: Gestation of the eggs takes up to four weeks at 75-82° F (24-28° C) (O. Hes pers. comm. 2021).

Bell Cricket

Meloimorpha japonica (Haan, 1844)

This is the only pet orthopteran with its own line of personal care products. This includes enclosures and decorations, food, jellies, substrate, and even water. The popular name in Japan is suzumushi. (Suzu means 'bell' while mushi refers to 'bug' or 'invertebrate,' as it is used for insects, crustaceans, and snails.) The bell cricket has been kept as a pet by the Japanese for 250 years and is commonly available at specialty pet shops. The species is also native to China and India. Adults are less than an inch long. The male's song is a pulsing, trilling song unmistakable as a cricket.

Bell cricket food packets

Life Cycle: Nymphs molt every week or two, a little slower than feeder crickets. The adults live only a few months. The big difference between *M. japonica* and other feeder crickets like *A. domesticus* is that the eggs take three to four months, not ten days, to hatch.

Husbandry: Bell crickets are commonly kept in solid-walled cages with vented lids. They are provided around 2″ (~5 cm) of substrate over the bottom of the entire enclosure to retain moisture and for egg laying. Like other crickets, dry dog food and fruit slices can be used as a primary diet. Specialty foods made just for suzumushi do not seem to be of much use for this species (Larsen 2019) or others. I purchased a small box of suzumushi food from Japan and it is not well-liked by house crickets.

Ant Crickets

Myrmecophilus spp. (Berthold, 1827)

These diminutive crickets are typically found around and inside of ant colonies. They eat mostly stray organic matter and ant refuse, though at least one species, *M. americanus*, has been observed tricking ants into feeding them via trophallaxis.

Bell cricket wood, jellies, water, food, and instruction sheet in Japanese © Kree McMonigle

These bell crickets are being sold in a Japanese pet shop in Tama City, Tokyo. If you look closely you can see two males singing. © Kree McMonigle

Eastern ant cricket, *Myrmecophilus pergandei,* in Virginia (CC-BY Judy Gallagher)

An ant cricket in Australia (CC-BY Jean and Fred Hort)

Myrmecophilus manni, large nymph, from Kuna, Idaho © Tristan Shanahan

Life Cycle: Eggs take a little over a month to hatch at room temperature (72° F or 22° C), and there is only one generation a year for U.S. native species.

Husbandry: These are very fragile little crickets. Only one species has been reproduced in captivity to date, *M. manni*. They feed and grow on a variety of grain-based foods, fruits, and possibly food molds. They seem to prefer cool temps 68-73 F° (20-23° C), and high moisture levels. Unfortunately, they cannot tolerate competing microfauna, especially mites and prolific springtails, making them difficult to culture for long.

Southern Mole Cricket

Neoscapteriscus borellii (Giglio-Tos 1894)

The southern mole cricket is a large (2.5-3.5 cm) species found from Florida north to North Carolina, west to Texas and in California. This and two other *Neoscapteriscus* were accidentally introduced to Florida around 1900 from South America. These neotropical species cannot survive further north, even underground. Although North America has several native and introduced mole crickets, the only time I found one was a male at a rest stop in Florida in 1988, dug up while it was singing. There were others singing, but I was lucky to catch one in the sand and roots. I have never seen mole crickets on a vendor list, but they are considered important lawn pests—they are not widespread enough for transport restrictions to be considered irrelevant. *Gryllotalpa gryllotalpa* is a widespread adventive in the eastern U.S. and widely across its native Europe, but the population in the U.K. is listed as endangered, so if you live there do not even consider touching one. They jump a little, as nymphs mostly, but hardly more than their body length.

Life Cycle: *Neoscapteriscus* reach adulthood in less than a year, but *Gryllotalpa* can take three years in nature (Haes

Mole cricket front legs are fantastically adapted for digging. The claw shape and placement can be different for each species.

Northern mole cricket, *Neocurtilla hexadactyla* (USGS Bee I&M Lab, Maryland)

Mole cricket, *Gryllotalpa* sp., Australia (CC-BY Jean and Fred Hort)

Neoscapteriscus borellii adult dug out of the sand next to a lake, Yulee, Florida

Neoscapteriscus borellii adults and nymphs can live together safely.

1992). Females lack an ovipositor and lay a pile of eggs in a chamber underground. This happens only once or twice—there are up to 300 ova for *Gryllotalpa vulgaris* (Stanek 1969), but *Neoscapteriscus* only lay 30.

Husbandry: Mole crickets can be kept on damp, sandy soil to allow for burrowing, 1-4″ (2.5-10 cm) deep. They require little ventilation, so a 2 x 2″ (5 x 5 cm) screened vent on the lid is plenty. More ventilation means more attention must be paid to maintain substrate dampness. They eat dog food, like most crickets, but leftovers must be removed because the dampness needed for solid burrow construction prevents a dry spot for food. In nature they feed on beetle larvae and earthworms in the soil, damaging plants mostly by undermining them. Ants are a primary food for *N. borellii* in riverine habitats in Brazil (Preston-Mafham 1992). In captivity different stages can be kept together. They do not cannibalize, not even dead, so adults should be removed as they die. Fruits and vegetables get little attention in an enclosure, though over a few weeks, carrot is slightly nibbled on. They are fantastic to look at and energetic like moles, but it is difficult to observe them even in a clear-sided, narrow enclosure. If they are not dug up now and then, the terrarium is akin to pet dirt, but digging them up during molting or ova gestation can result in damage. It is surprising how many can fit in a handful of substrate, totally invisible until you dig through a few times over. Mole cricket males use tunnel construction to amplify their singing, but in captivity this may be more difficult. One must be within a foot or two of the enclosure to hear the gentle singing of *N. borellii*.

Zambian Cave Cricket

Phaeophilacris bredoides Kaltenbach, 1986

This popular pet cricket lives in Zambia and bordering areas of southern Africa in the wild. Founding specimens were supposedly collected in a karst cave by the name of Chibongwe, but it is not found on any maps. In captivity, all specimens originate from a man named Ludek Dobroruka who collected them in 1968 and brought them back to Prague, Czech Republic. The crickets were traded around insectariums and museums under various incorrect names from 1968-1986 when they were described as a new species (Larsen 2002). As of 2021 the original stock remains stable and widespread. The eyes are normally dark brown to black but there are mutants with lemon-yellow colored eyes.

Life Cycle: Adults, like many other crickets, have a short courtship with some wing flapping and antennae tapping, ending up with the female on top. After mating the female produces up to 200 eggs, a few at a time, over about a five-month period. They are laid sporadically, but a large number can be laid in one night if there is no substrate for laying, but a good media for laying is then introduced. Eggs hatch after about five weeks when kept at room temperature (72° F or 22° C), but this can be slower if kept cooler or faster if kept 10% warmer. Nymph development also varies greatly with temperature. At 68° F (20° C), they mature in ten months—six months at 86° C (30° C) (Bruins 1999).

Husbandry: The cage should have a screen lid and a few inches of damp soil. There is no need to keep a dry area. Cork bark planks are normally used in place of egg carton for molting and perching because they are less likely to mold or

Phaeophilacris bredoides, mature male with yellow eyes, 2011

Phaeophilacris bredoides, male with standard eyes © Tom Larsen 2001

disintegrate. Zambian cave crickets have proven more of a display animal or pet cricket and attempts to use it as a feeder of real quantity like *Gryllus* will quickly fail. Also, they jump many times higher than house crickets so handling and containment are more challenging for feeder use. Eggs are laid in damp, sandy soil, a few at a time, and the hatch rate is good if they do not dry out. Like house crickets, the adults dig up eggs and eat them, which can greatly reduce numbers. Buried floral foam or fine netting laid over the substrate can reduce this problem if large numbers are desired. Nymphs take six to ten months to mature, with males and specimens kept at warmer temperatures maturing closer to the shortest amount of time. These crickets appreciate artificial cave structures and love to perch upside-down. It is not necessary, but it makes a great display.

Phaeophilacris bredoides nymphs are easy to sex. The female, upper right, is broader and has a rudimentary ovipositor.

Gryllus pennsylvanicus

A tiny cricket in Australia

Broad-winged tree cricket
Oecanthus latipennis
Virginia

Endacusta sp., spider cricket
Australia

Ground cricket
Neonemobius sp.
Alabama
(CC-BY Judy Gallagher)

Japanese burrowing cricket
Velarifictorus micado
Virginia
(CC-BY Judy Gallagher)

Robust ground cricket
Allonemobius sp.
New York
(CC-BY Judy Gallagher)

This large nymph *Nisitrus* sp. was found on Mount Kinabalu in Borneo. *Nisitrus* are day-active and often brightly colored as adults. © Ondřej Hes

True or cave cricket? Kruger National Park, South Africa © Ondřej Hes

Black-sided camel cricket
Ceuthophilus latens
Penitentiary Glen, Kirtland, Ohio

Camel cricket
Ceuthophilus sp.
Washington D.C.
(CC-BY Katja Schulz)

Arizona camel cricket
Pristoceuthophilus arizonae
Chiricahua National Monument, Arizona
(CC-BY Katja Schulz)

The camel cricket cave is often tiny and wooden in nature. Pipestem, West Virginia

Ceuthophilus agassizii
Adult female
© Tristan Shanahan

These large *C. g. gracilipes* nymphs are easy to tell apart from adults, not by size, but because they still have the immature color pattern.

Cave Crickets

Cave crickets are wingless, so the males do not sing. The antennae are strikingly long, and they can jump incredible distances. They hail from the cave cricket superfamily Rhaphidophoroidea and are not true crickets, but they are more like a cricket than other orthopterans. Habitats include caves, but also hollow logs and stumps, rodent and turtle burrows, and man's creations from hothouses to water boxes. These include the *Macrobaenetes* sand treader crickets from California and the huge *Tropidischia* found from California north through B.C. Canada to Alaska.

Scrubland Camel Cricket

Ceuthophilus agassizii (Scudder, 1861)

This pale cricket is found in scrub areas of the northwestern United States. They can be found in caves, but the majority are found in fissures in rock faces and under stones or wood on the ground.

Life Cycle: Eggs are usually the overwintering stage in the coldest temperate areas (some northern species and strains may require diapauses as nymphs or as eggs in captivity), whereas adults and nymphs may be found under logs or stones year-round in moderate temperate areas. At room temperature (72° F or 22° C) the eggs normally take six to eight weeks to hatch, and they reach maturity in three to four months.

Husbandry: The scrubland camel cricket is very easy to breed consistently, partly because specimens are more tolerant of crowding than other camel crickets. It is a terrestrial species, with a bulky build and short legs relative to the forest dwelling, arboreal camel crickets. Specimens should be offered plenty of hides in the form of egg crate or bark. Despite being found in arid habitats, this species prefers high humidity. As is the norm for camel crickets, all stages are protein hungry and should be fed dog food or some equivalent as their staple diet to prevent cannibalism. Fruits and veggies including carrots, peas, and corn can be offered supplementally. One problem encountered when breeding these and other species of camel crickets is substrate fouling. If the substrate gets too filthy with frass and leftover food crumbs (which camel crickets leave behind in abundance), their thin-shelled eggs rot and die in the substrate due to pest problems, mold, or harmful bacterial buildups.

Slender-Legged Camel Cricket

Ceuthophilus gracilipes (Haldemann, 1850)

This species is one of the largest, if not the largest, U.S. native camel cricket species. (The only other contender would be *Tropidischia xanthostoma*, which may have a longer leg span, but in terms of bulk *C. gracilipes* emerges the winner.) There are two subspecies, *C. g. gracilipes*, which is widespread across the east coast, and *C. g. apalachicolae*, which is restricted to the Apalachicola National Forest in Florida.

Life Cycle: The slender-legged camel cricket is an arboreal species, typically found resting on tree trunks at night in wooded areas. The legs and antennae are very long compared to ground-dwelling *Ceuthophilus* spp., and they can jump much further as a result. At room temperature (72° F or 22° C), eggs take up to two months to hatch, but development time is shortened quite a bit if kept in the 75-80° F (24-27° C) range.

A *Ceuthophilus* cave cricket is molting in 1996. Its parents were collected from under a rotten log in Hinckley, Ohio, 1995.

Adult male *Ceuthophilus g. gracilipes* © Tristan Shanahan

Adult female
Ceuthophilus g. gracilipes

Ceuthophilus g. gracilipes hatchling
© Tristan Shanahan

Husbandry: A screen-topped cage with an entry port is preferred to allow replacement of uneaten dog food that could mold. (Water can just be poured through the screen onto the substrate.) Some large pieces of bark should be leaned against the sides of the cage so they can hide underneath, with smaller bark pieces or other small shelters for young nymphs available also. This species does not tolerate crowding well and should be given lots of space, otherwise cannibalism will ensue, and large molting individuals will often be knocked down from perches, mid-molt, by their tankmates, resulting in death. One thing that these and all relatives seem to be capable of is teleportation. If a keeper is close enough to disturb them, they can jump past, out of the line of sight, so quickly they seem to disappear. It is not a huge challenge to husbandry if the keeper is careful and aware. Still, the spectacular jumping range, along with the crazy long antennae and legs, is probably the reason they scare those who lack an appreciation for insects.

Ceuthophilus g. gracilipes ova
1.5 mm x 3 mm

Greenhouse Camel Cricket

Tachycines asynamorus (Adelung, 1902)

This Asian adventive seems to be more commonly encountered than the native *Ceuthophilus* in North America, though it may simply be better adapted to disturbed habitats around humans. I collected some under the walkway around my sister-in-law's pool in Norfolk, Virginia, in 2006. In 2013 I saw three in a glasshouse in Cleveland, Ohio, but they were twitching as they convulsed from pesticides. Their spidery appearance and fantastic jumping skills make them a target of fear, but they are a beauty to the orthopteran lover. Greenhouse camel crickets are common around pools, sheds, and greenhouses. They do not cause any real harm—but some humans lack appreciation for their finer beauty. They mostly feed on other insects, but may chew on seedlings and flowers at times.

Life Cycle: At room temperature (72° F or 22° C) ova take six to eight weeks to hatch. Nymphs grow slowly, taking five to seven months to mature and adults live another six months. At greenhouse temperatures the cycle is probably faster.

Husbandry: As with most every cricket, they eat kibbled foods for dogs and cats, fish food pellets, various fruits and vegetables, dead insects, and almost any meat-based organic item they can scavenge. They will nibble on green leaves and rotten wood when hungry, but it is unlikely nymphs would survive and grow on these. They love to drink milk from a small dish. Eggs are laid individually in damp soil about 1/2″ (1.2 cm) deep. They should be kept damp throughout gestation.

Tachycines asynamorus adult male in Norfolk, Virginia, 2006

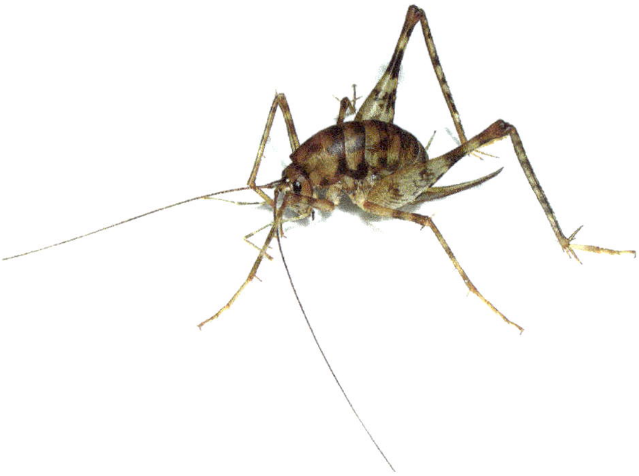

Tachycines asynamorus adult female, two generations removed, 2008

An average-looking, tropical cave cricket in
Endau Rompin National Park, Malaysia © Ondřej Hes

Hadenoecus subterraneus, Mammoth Cave National Park, Kentucky
(CC-BY James St. John)

A female king cricket, *Libanasidus vittatus,* wanders Kruger National Park, South Africa. This common species is known as the Parktown Prawn. The adult males have huge tusks for combat. © Ondřej Hes

Cook Strait giant weta, *Deinacrida rugosa* (CC-BY Sid Mosdell)

King Crickets

Members of the Superfamily Stenopelmatoidea are known as wetas in New Zealand, king crickets in Australia and South Africa, Jerusalem crickets in the United States, and 'children of the earth' in Mexico. Most species are wingless giants. Males of some species have massive jaws and females can have long or short ovipositors.

Jerusalem Crickets

Ammopelmatus spp., (Tinkham, 1965) & *Stenopelmatus* spp., (Burmeister, 1838)
For the longest time, all but two of the U.S. Jerusalem cricket species were classified under the genus *Stenopelmatus*, but in a recent revision (Weissman, 2021), all U.S. species were moved to *Ammopelmatus*, with true *Stenopelmatus* only being found south of the U.S. border. These crickets have long been brought into the hobby as novelty pets, but there has been little to no success breeding them.

Life Cycle: Nymphs take one to two years to mature depending on species and natural diapause. The adults live 6-8 months at most, often less if breeding. Eggs likely only take a couple months to hatch for species that lay during the spring and summer, longer for those that lay later in the year in colder climes, where their eggs must overwinter.

Husbandry: Nymphs and adults are best maintained at room temperature (72° F or 22° C), though temps 75° F (24° C) or higher may be needed for adults of some southern species to reproduce. Northern species may require a winter diapause. The best basic substrate choice seems to be coconut fiber—most keepers who use just sand or add sand to their mixes report poor success in keeping the animals alive for long. It seems too abrasive for them (with the exception of the few dune-dwelling species). They spend much of their time underground, and so should be given a substrate several inches deep. These are very protein-hungry orthopterans, and *Ammopelmatus* do best on a staple diet of dog food and live invertebrate prey, with fruits and veggies as supplements, whereas *Stenopelmatus* appear to be strict carnivores that may not even nibble on dog food.

Keeping them communally is out of the question, and breeding pairs must be put in temporary mating containers, with very shallow, smoothed out substrate.

Ammopelmatus pictus female,
West Stanislaus County, California
© Tristan Shanahan

Ammopelmatus sp. from Long Beach, California © Tristan Shanahan

Courtship consists of males drumming their abdomens against the ground in species-specific patterns, with receptive females drumming back. The male then 'attacks' the female and bites her hind legs and ventral abdomen until he positions her properly for copulation. Immediately after mating the male should be removed from the container. The female should be left there overnight before being put in a container with deeper substrate, because if allowed to burrow right away, she can rub the fresh spermatophore off in captivity.

The greatest barrier to breeding Jerusalem crickets is inducing oviposition and incubating the eggs. It has proven quite difficult to get mated females to lay eggs in captivity. While evidence suggests females lay all their 30-40 eggs at once in an underground chamber, enthusiasts who have collected eggs in captivity usually only find a few eggs, laid singly in the substrate. Considering that eggs removed from the abdomens of females that died seemingly from being egg bound (a common problem for those breeding Jerusalem crickets) have all appeared to be fully developed, it would seem the eggs laid singly are an artifact of incorrect husbandry, not the actual norm. Eggs are likely supposed to be laid in one go (hence the females getting egg bound without proper oviposition impetus). A clay-based lower substrate may be needed for inducing females to create egg chambers and oviposit properly. So far it seems that the eggs are sensitive to over-watering, and that they do not like being covered completely in substrate (a fact lending itself to the egg chamber theory as well). Eggs from the Asian *Sia* spp. are also teardrop-shaped, but pale green rather than yellow.

Wellington Tree Weta

Hemideina crassidens (Blanchard, 1851)

This type of New Zealand weta is not threatened, but some relatives, primarily the large ground wetas (*Deinacrida* spp.), are the subject of conservation efforts and intense regulations. The ground wetas are particularly threatened by introduced rodents, so indirect conservation efforts (anything other than protected breeding) will fail unless a method to eradicate or control rodents is discovered. 10 of the 11 *Deinacrida* are protected by law.

On the other hand, tree wetas (*Hemideina* spp.) are easily farmed and are primarily threatened by man's landscaping and development efforts. They may be encouraged in the same way as providing bird houses for bluebirds. Tree wetas usually spend the day in holes in rotten logs hollowed out by other creatures or decay. Unfortunately these homes are normally cleaned up around human habitations and farms. Artificial living chambers can be purchased and placed in acceptable habitats. The male's huge head and jaws are used to keep other males out of the chosen living quarters. A few specimens of this particularly handsome species arrived in Europe in 2001, though the few that survived shipping did not live long—to date, Wellington tree wetas are only available as dried specimens.

A female giant king cricket, *Anostostoma australasiae*, Australia (CC-BY-SA Bernard Dupont)

Male tree weta holding a pencil in its jaws, 2001 © T. Larsen

Male Wellington tree weta, *Hemideina crassidens,* 2001 © T. Larsen

Adult female weta from New Zealand, 2001 © T. Larsen

Huge jaws and the ears on the front legs are notable from this view.
Wellington tree weta, *Hemideina crassidens,* 2001 © T. Larsen

Large nymph of *Hemideina crassidens,* 2001 © T. Larsen

Mercury Islands tusked weta
Motuweta isolata
(CC-BY B. Robertson / NZ DOC)

Spizaphilus sp.
Madagascar
(CC-BY-SA Pavel Kirillov)

King cricket
Queensland, Australia
(CC-BY Greg Schechter)

Jerusalem cricket
(CC-BY John E. Robertson)

Leaf-rolling cricket
Family Gryllacrididae
Cuivre River State Park, Missouri
(CC-BY Andy Reago and Chrissy McClarren)

Cooloola monster
Cooloola propator
Queensland, Australia
(CC-BY CSIRO ScienceImage)

Acanthoplus discoidalis were seen at a European fair, 2007. Wild-caught *Acanthoplus* have shown up sporadically for decades, but have not reproduced. © Tom Larsen

Long-legged armored katydid, *Acanthoplus longipes,* in terrarium, 2001 © Tom Larsen

Katydids

The Tettigoniidae are also called 'long-horned grasshoppers' or 'bush crickets', which can be very confusing since the qualifying 'long-horned' and 'bush' are often dropped off by writers and speakers. Katydids are mostly day-active but sing towards dusk and may come to lights at night. Many species look like green leaves, others like armored monsters, and many more appear as obvious predators with spines on the front legs or all the legs to hold prey like a thorny basket.

Amblycorypha rotundifolia: rare yellow-mottled form, Trumbull County, Ohio

Oblong-Winged Katydid

Amblycorypha oblongifolia (De Geer, 1773)

This leafy green katydid is widespread across the eastern United States and southeastern Canada. There are several similar-looking katydids, all around 2″ (5 cm) including the wings, in the same areas from various genera, but this one is particularly interesting for zoological displays because odd color forms (notably pink and orange) are commonly encountered and can be isolated in captive cultures. Unnatural colors may sabotage crypsis in the wild, but are spectacular in captivity. Adults, usually males, fly to lights at night in late summer and stay around through the next morning.

Life Cycle: Eggs are laid in soil and usually hatch the following spring. If kept cool (22° C or 72° F), or lower, through late spring, a portion of the eggs will wait till the following spring to hatch even if temperatures increase later. Nymphs reach maturity after about four months. Females lay eggs and die around the time the hard frosts begin, even if protected from the cold.

Husbandry: A screen cage is recommended. Oblong-winged katydids feed on a variety of vegetable and animal material, but mostly on fresh leaves. Captive specimens can be reared entirely on blackberry or rose leaves. They seldom cannibalize unless they are crowded, or if leaves are wilted or mostly eaten. For this reason, enclosures should provide at least 3 gallons (11.4 L) of space per animal. Cannibalism, at worst, is usually limited to specimens eaten in the middle of molting or wing edges of adults nibbled on as though another specimen thought it might be a leaf.

Malaysian Leaf Katydid

Ancylecha fenestrata (Fabricius, 1793)

Source specimens for the commonly kept stock were collected in 2003 in the Tapah Hills of Perak State in northwestern Malaysia. Adults are plus or minus three inches long including the wings—males to 7 cm and females to 8 cm. They have

Amblycorypha oblongifolia orange and pink forms at the Audubon Butterfly Garden and Insectarium, January 2014 © Dan McMonigle

Amblycorypha sp., Virginia (CC-BY Judy Gallagher)

Ancylecha fenestrata male, December 2010 © Tom Larsen

Ancylecha fenestrata subadult male © Tom Larsen

Ancylecha fenestrata, unusual yellow male, 2010 © Tom Larsen

Ancylecha fenestrata subadult female, October 2010 © Tom Larsen

a standard pattern of white-edged, black crescents on the outside of the green tegmina which sets them apart from most leaf katydids. Another handsome feature is that the wing pads of large nymphs stick up straight off the back like shark fins. Males and females both chirp in defense. This is one of the duetting katydids where female chirps respond to the male's singing and aid them in location.

Life Cycle: Eggs are inserted sideways into the pith of leaves, as many as 13 in a leaf. Plants used for this include privet (*Ligustrum*), which they also eat, and two common houseplants: English ivy (*Hedera helix*) and spotted Dracaena (*Dracaena surculosa*). The young hatch after 70 days at 72-77° F (22-25° C) (Grevin et al. 2013). Nymphs are very flat when they first hatch but quickly inflate. They take five months to mature and then live six months as adults.

Husbandry: A terrarium 40 x 35 x 60 cm (16 x 14 x 24″) is perfect for three to four pairs. This should be kept at 75-90% humidity, with a suggested temperature between 77-83° F (25-28° C), a few degrees cooler at night. The favorite food is smoke tree (*Cotinus*), but they also feed on the leaves of butterfly bush (*Buddleja*), forsythia, sumac (*Rhus*), St. John's wort (*Hypericum perforatum*), bramble, privet (*Ligustrum*), and apple slices. Overcrowding may lead to cannibalism, but adequate space prevents this problem (Larsen 2011a).

Black-legged meadow katydid, *Orchelimum nigripes,* my favorite indoor singer

Green and brown forms of *Arachnacris corporalis*
at the Iowa State Insect Zoo © J. Byrne

Giant Malayan Leaf Katydid

Arachnacris corporalis (Karny, 1924)
This is a large, gentle, flower eater from Malaysia and Sarawak. Adults can reach nearly 6″ (15 cm). Huge size, ease of care, and availability from regular zoo suppliers made it an occasional display animal at insect zoos in the past. Only over the last decade did keepers learn how to induce egg laying and begin breeding them through generations. The male's call is very loud, and its husbandry may preclude those with both a domiciliary lab and life partner.

Life Cycle: The massive females lay eggs in crevices where the bases of leaves overlap on the surface of plants. Commonly used plants are bromeliads (a type of plant mostly restricted to the New World) and bananas. This is probably not ideal, but they have been observed laying eggs in soaked chunks of wood at Iowa State Insect Zoo (Byrne pers. comm.) so floral foam may work well. The survival rate is good, and nymphs are not known to cannibalize.

Husbandry: A full screen cage is suggested, but a large glass cage with big

Arachnacris corporalis adults at the Smithsonian insect zoo in Washington D.C., 2012

Yellow form *Arachnacris corporalis* at the Iowa State Insect Zoo © J. Byrne

Female (top) and male (bottom) *Capnobates fuliginosus* from Madera Canyon in Arizona

branches will be successful. Flowers such as daylilies are a favorite treat, but they feed on a variety of common produce from the grocery. Staple foods include romaine lettuce, apple slices, and sweet potato chunks. In zoological displays the foods are normally offered on a wooden skewer for easy removal of the leftovers. Normal developmental temperatures range from 75-80° F (24-27° C).

Sooty Longwing Katydid

Capnobotes fuliginosus (Thomas, 1872)
In terms of wing length, the longwings are among the biggest North American species, with a head to wing tip measurement of over 3″ (8 cm). The body is relatively small (<1.75″ or 4-4.5 cm) from head to tip of the abdomen, not including an inch for the female's ovipositor. The sooty longwing is a common species in Arizona, and ranges north to Colorado and west to California. Other southwestern *Capnobotes* have green, brown, or gray forms but this species is normally only gray. The common name mostly refers to the gray hindwings which are held two-thirds open in the threat display. It rests against the huge wings and slightly pulses them. This display makes the animal look very large and somewhat menacing. The longwing is an excellent flyer and readily jumps and flies away. The black wings are flashed in flight and then closed as soon as it lands. It is one of the few orthopterans capable of using the wings in both a deimatic and startle display.

Husbandry: This species may eat various leaves and vegetables, but live house crickets are an easy staple diet in captivity. They usually hunt in the evening, so if there is no ambient lighting after the main lights go out, they may not hunt. Only one feeder cricket should be offered at a time to prevent a reversal of the predatory roles. Longwings should be kept in individual enclosures except for mating.

Rhinoceros Katydid

Copiphora rhinoceros (Pictet, 1888)
With its robust yellow horn, thorn-studded legs, and translucent neon green body, this medium-sized katydid is striking to behold. Like the dragon-headed katydid, it rarely actively hunts prey, but feeds on almost anything it comes across. In nature it has been observed feeding on fruit, snails, frog eggs, other katydids, and small lizards. The horn is said to be used in defense, but I have never seen it purposely wielded (Byrne 2021). However, it makes swallowing this insect a painful experience for many predators. The massive jaws are more of a threat to prey and handlers. It is difficult to handle a rhinoceros katydid without being bitten at least once. The male alternates chirping and shaking leaves to attract the females.

Male *Copiphora rhinoceros* at World of the Insect, 2021

Female *Copiphora rhinoceros* at the Smithsonian Insect Zoo, 2010

Copiphora rhinoceros nymph at the Iowa State Insect Zoo © J. Byrne

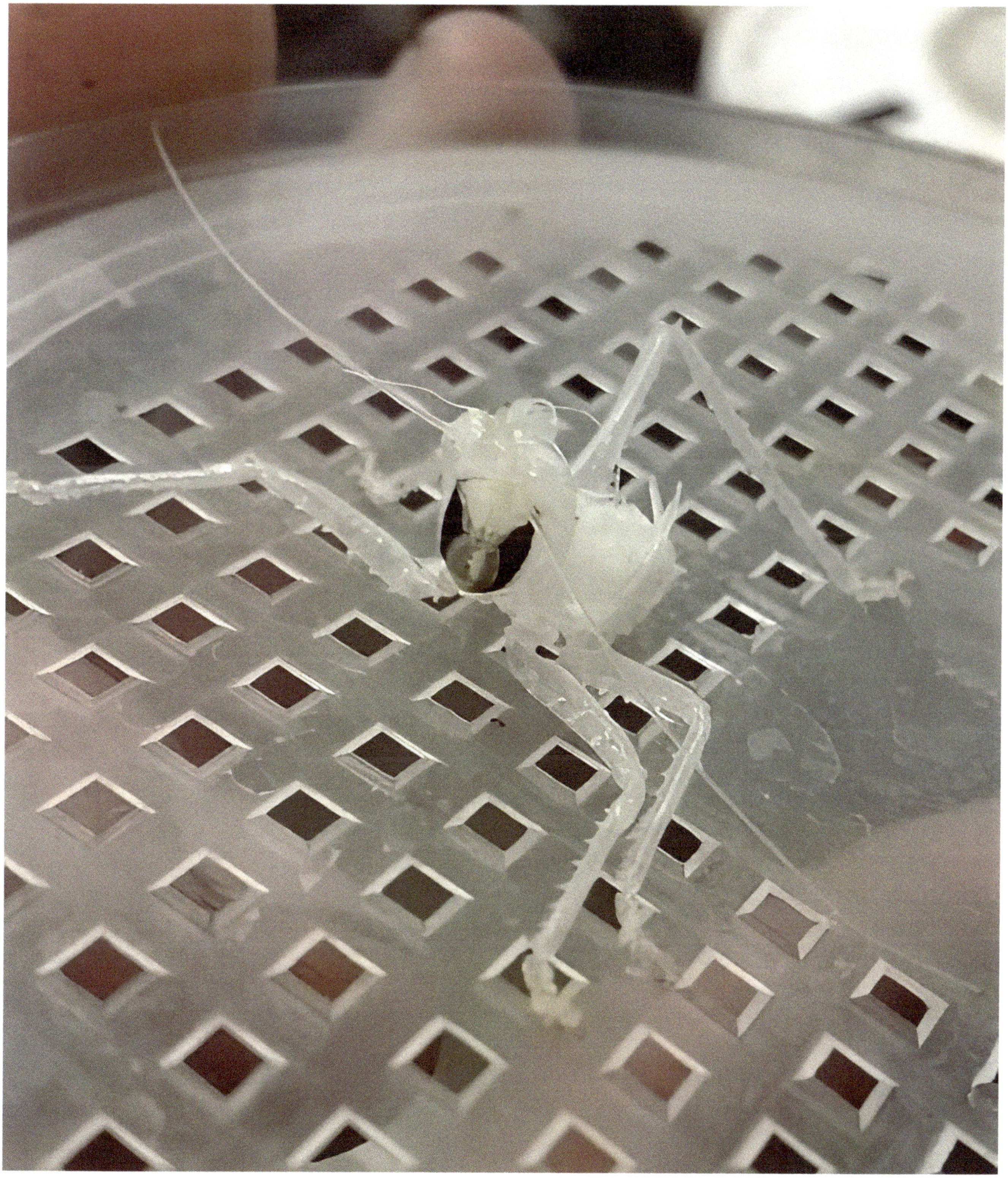

Copiphora rhinoceros shed at the Iowa State Insect Zoo © J. Byrne

Copiphora rhinoceros, small nymph,
Iowa State Insect Zoo © J. Byrne

Life Cycle: Female *C. rhinoceros* have a sword-like ovipositor longer than their body, and can use it to insert eggs either in soil or plant tissues. If kept warm and moist, the eggs will hatch in approximately four months. The males attract females with a chattering call and body-shaking display, and like most katydids produce exceedingly large spermatophores and nutritious spermatophylaxes for females to feed on. The huge sperm packet produced by the male measures 4/5 x 2/5″ (2 x 1 cm) (Preston-Mafham 1992). This is a long-lived katydid; breeding males have lived for ten months as adults at the Iowa State Insect Zoo (J. Byrne pers. comm.).

Husbandry: Crunchy and juicy foods are particularly appealing to this katydid, which can be fed a staple diet of apple, carrot, romaine lettuce, and cockroaches. Other tempting foods include banana, beetle jelly, and waxworms. Presenting food on a skewer works well, as this katydid commonly rests near the top of its enclosure or on broad leaves and prefers for prey to come to it (though it will travel around the enclosure at night). *Copiphora* have no qualms about cannibalism, and nymphs should be kept separate to ensure maximum survival. Nymphs are also prone to molting difficulties, even when kept at very high humidity. Ensuring that the katydid has water droplets to drink and a variety of angled surfaces on which to molt can help reduce losses.

Dragon-Headed Katydid
Lesina intermedia (Karny, 1923)
This popular and fantastic insect zoo inhabitant has been imported as display specimens from Malaysian suppliers since at least the late 1980s. However, little effort was put into captive reproduction, partly because of uncertain egg-laying requirements. Females seemed to eat well

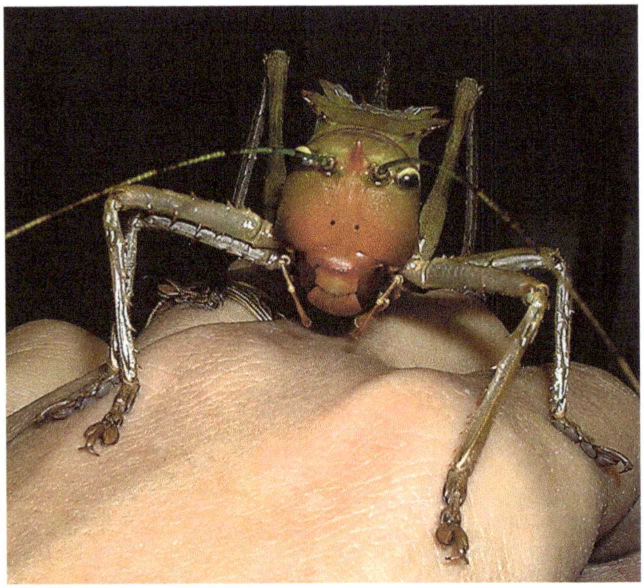

Head shot of *Lesina intermedia*,
behind the scenes at the Cleveland
Metroparks Rainforest, 2004

Horned katydid in Los Tuxtlas, Mexico © Ondřej Hes

Lesina intermedia from a Malaysian butterfly farm displayed at the Cleveland Rainforest Insectarium, 2004

A female *Lesina intermedia* cleans her ovipositor after mating.
World of the Insect, 2021

A female *Lesina intermedia* inserts eggs into a banana tree pseudostem.
World of the Insect, 2021

and lived up to a year, but never laid any eggs. They were simply purchased for zoo display, like cut flowers. I had a chance to handle an adult female within a permitted local containment unit a few decades ago and they were spectacular. I am usually not fearful of orthoptera, but I was ready for her to just take a big chunk out of my flesh, though she did not show aggression.

Life Cycle: A more recent zoo display katydid (*Arachnacris*) was being reared in captivity in the late 2000s and its egg-laying preferences gave way to the discovery of what to do with this species. Chang (2014) found females readily insert eggs deep inside the soft, pseudostem ('trunk') of a banana tree. Eggs moved to incubators had a reasonable hatch rate, but more than half died when removed from the pseudostem. Kuar et al. (2018) found the hatch rate of *Lesina* (*Eumegalodon*) *blanchardi* averaged 96.3% if ova remained inside the pseudostem until hatching. Other plants with large fleshy pseudo-stems and leaves like bird-of-paradise and fan palm should also work. Adults are hardy but young have a low survival rate, especially when reared in groups. Imported adult specimens commonly live nine months or more (Cleveland Zoo keepers, pers. comm., 2004).

Husbandry: Dragon-headed katydids eat freshly killed crickets and their siblings, but they are opportunistic predators or scavengers and seldom hunt actively. They feed on a wide range of vegetable materials including romaine lettuce, bramble, cucumber, zucchini, sunflower, and pumpkin seeds (shelled). Like many orthopterans, they will feed on dry dog food and other kibbled foods for predators, or fish pellets, but not pelleted rodent foods. Specimens should be kept in separate cages except for mating. The temperature should be 75-80° F (24-27° C) for gestation and development, though adults can survive much cooler temperatures. Humidity should be at least 60%.

Angle-Winged Katydid

Microcentrum rhombifolium (Saussure, 1859)
Like *Amblycorypha,* this species and others in its genus often dwell in treetops. Adults may be attracted to lights, while the sturdy-bodied nymphs are sometimes found by sweep-netting saplings or low branches of maples and oaks. Unlike *Amblycorypha*, all individuals are some form of green with subtle white mottling. Angle-wings are some of the largest katydids in the northern U.S. and very easy to raise. Rather than singing incessantly, males may also perform a wing-flapping display when females are nearby.

Microcentrum nymphs in 1996, reared on miniature rose

Life Cycle: Females lay their scale-like eggs in overlapping rows on twigs or leaf edges. The eggs may hatch from March to June throughout most of the U.S.—in the extreme south, *Microcentrum* can be found year-round. A nymph takes around two months to mature, and the adults live another two or three months.

Husbandry: *Microcentrum* does well in a full-screen cage, though a tall well-ventilated terrarium can also be used. The leaves and flowers of a wide variety of plants are accepted. Rose, oak, maple, hibiscus, grapevine, and bramble are reliable choices. This is a hardy katydid which rarely cannibalizes or suffers bad molts.

Red-Eyed Devil Katydid

Neobarrettia spinosa (Caudell, 1907)

This vicious monster is a sought-after exhibit specimen and is possibly the most spectacular North America orthopteran. The eyes are red, while the body is bright green and white. Also known as the greater arid land katydid, *N. spinosa* is an aggressive, obligate predator. The gigantic, spiny legs make it appear even larger than it is. The adult male has an impressively loud mating call which is sung during the day in captivity.

The powerful jaws are readily used in defense and the flexible body makes it nearly impossible to hold a specimen in such a way that it cannot bite. Adults can jump up to a meter. Its threat display includes rearing up its body, showing open jaws and spread wings. The rear (flying) wings used in the deimatic display are black with blue spots. They are reserved for this purpose as they are too short for use in flight. The threat display of large orthopterans is different from the startle display of many small species that quickly reveal, then hide, bright colors in hopes of escaping. These stand their ground.

Wild specimens are often found at the tops of bushes or in trees, especially densely-foliated trees such as junipers. In nature they feed on other orthopterans, stick insects, mantids, and small vertebrates like lizards and nestling birds. *Neobarrettia* may be found in suburban areas, but reaches its greatest density in scrub forests and semi-desert areas.

Life Cycle: The mature female has an immense, sword-like ovipositor and a larger abdomen than the male—the head, legs, and thorax of the two genders average the same size. The ovipositor is short and difficult to see on hatchlings, but enlarges with each molt and is obvious by the third instar. Male nymphs look similar but have no ovipositor; upon maturity their outer wings (tegmina) look rather different since they are developed for stridulation. Females drive the ovipositor deep into the substrate to lay an egg. and usually lay two to ten eggs each day. Damp, mostly sand substrate can be placed in a 32 oz. deli-cup for oviposition. Eggs are tan cylinders about 1/3″ (8.5 mm) long and 1/16″ (1.5 mm) wide.

The greatest barrier to successful reproduction of this species is hatching the eggs. If kept above 85° F (29° C) they could possibly hatch in three months, but in my experience, it took more than two years with alternating hot 92° F (33° C) and cool 65° F (17° C) cycles to finally get them to hatch. Despite experience, the second attempt was no quicker. Ova should be kept damp the entire time, but are resistant to some drying. Fertile eggs contain clear liquid during most of gestation. Nymphs

Neobarrettia spinosa, female, from B. Tomberlin, Hatari Invertebrates, 1998, on the hand of the late James V. Smolka, entomologist

A captive-hatched (2004) immature *Neobarrettia spinosa*, shows off its nymphal threat display. Notice the spines on all the legs used to capture and hold prey in a deadly embrace.

can only be seen weeks before hatching upon dissection. Opening an egg here or there to determine fertility of a group of eggs laid by the same female is unnecessary and will result in unacceptable losses over the long gestation. Healthy eggs will remain plump while those that die or were never fertile deflate and begin to disintegrate.

Growth can be extremely rapid for such a large insect, with hatchlings reaching adulthood in as little as two months in nature. The captive growth rate can be significantly slower if temperatures are lower and the katydid is not constantly fed. Wild adults usually live three to four months, with males dying first, but some can live much longer, especially if they

Light-colored eggs from *Neobarrettia spinosa;* larger brown eggs from *Romalea microptera,* 1999

were reared in captivity and never mated. My first attempt ended with only three adult males and they lived about twice the usual time.

Husbandry: Upon hatching the small nymphs are hungry and must eat within the first few days or die. Depending on the length of gestation, they may be too weak to catch crickets. On day one, provision of a tiny slice of banana will give them the energy to hunt. The next day they can be given a cricket from 1/4 to 1/2 their mass. They will also scavenge pre-killed insects, especially if those insects are pinned to the wall of the enclosure or offered slowly with tongs. A primary diet of crickets works well. In captivity all stages catch and eat crickets and cockroaches, but will also chew on dog food pellets, banana, and some other foods. For successful development, the diet should consist almost entirely of live food. Specimens may mature at a greatly reduced size if prey is irregularly provided or fruit is commonly offered to stretch out feeding of live insects.

Molting is another tricky aspect of keeping this species. Small specimens, for the first few instars, have no trouble molting off the smooth sides of a plastic or glass container. But as they grow larger, they still need to hang upside down to molt and their great weight makes them fall from small or smooth surfaces. For molting the lid should be metal window screen (fiberglass screen can be chewed through) vented, and at least 4.5″ (11.4 cm) in diameter. Paper-vented lids and mulin cloth vents work fine for the fourth and fifth molt, but are too fine for the claws of late instar nymphs. Tall (6″ or 15 cm high) deli-cups are useful individual enclosures through subadult, but nymphs must be moved to larger containers for the final molt.

Although both genders are strong predators, the primary risk of cannibalism arises from doomed attempts at keeping immature specimens together. If given space for the introduction and one specimen of each sex is introduced, losses during mating attempts are rare. The male can usually be kept with the female for a few days in a large cage (30 gallons or 114 L) without risk of death.

American Shieldback Katydid

Pediodectes haldemanii (Girard, 1854)

This large, predatory katydid is easy to handle compared to *Neobarrettia*, but can still bite or jump away. Adults may be fun to handle, but if held in a closed hand they can give a nasty bite. This shieldback is found in Texas and bordering states in the U.S. Wild specimens are usually bright green. Adults cannot change color, but captive conditions usually lead to pale brown animals if they are reared from

Pediodectes bruneri

Pediodectes sp.
Weslaco, Texas
(CC-BY Judy Gallagher)

Pediodectes haldemanii
Ovipositing female

Pediodectes haldemanii compared to a man's forearm, Hatari Invertebrates, June 1999.

Pediodectes bruneri (from Texas, 2021) does not catch crickets, but readily feeds on adult western short-horned walking sticks, *Parabacillus hesperus*.

Pediodectes haldemanii
Texas, 2021

Pediodectes nigromarginatus
Female
Atascosa County, Texas

eggs or nymphs. Other *Pediodectes* like *P. nigromarginatus* are normally gray. The male's tiny wings are mostly hidden by the pronotum shield and make a short chirp.

Life Cycle: Females lay cylindrical eggs (1/5″ or 5.5 mm long by 1.5 mm wide) a little over an inch (~3 cm) deep in the soil to protect them from drying out. Eggs take five or six months to hatch at room temperature (72° F or 22° C). Adults live three to four months and nymphs take up to five months to mature.

Husbandry: An enclosure with a small access door should be used since the defensive jumps otherwise ensure escape. This large and aggressive katydid does not eat leaves of the normal katydid fare in captivity, but it will nibble on various types of sliced fruit. It may catch crickets sometimes, but freshly killed insects are more likely to be eaten. Specimens feed on dry dog food in captivity more than other common foods. If fed only dry dog food, mist them daily to provide drinking water.

True Katydid

Pterophylla camellifolia (Fabricius, 1775)
This is a large, beautiful species that is easy to rear from nymphs. Finding stock is difficult because it lives high in trees and seldom comes to lights as adults. (Many of the other types of leafy, green katydids can be found at lights in late summer, or found on bushes within reach of a butterfly net).

Life Cycle: Males call females from the tops of trees and females lay eggs under loose bark. Eggs hatch in the spring and adults die with hard frosts in late fall to early winter.

Pterophylla camellifolia female with visible ovipositor (short on this species and not always visible in photographs)

Husbandry: A full screen cage is suggested. If food is fresh and humidity is above 50%, daily misting is not necessary. They feed on leaves of oak and other tall trees in nature, but are easily kept on leaves of low bushes that other species eat, including blackberry, raspberry, and rose. The specimen here was reared on leatherleaf viburnum (*Viburnum rhytidophyllum*).

Giant Katydid

Stilpnochlora couloniana (Saussure, 1861)
This heavy-bodied species is native to Florida, Cuba, and various Caribbean Islands. The 'giant katydid' has been commonly available from European breeders in the U.K. since at least 2010 under this name. It is a big, bulky katydid, but 'giant' may be a bit of a misnomer as the length from the head to the end of the wings rarely exceeds 7 cm (less than 3″). Specimens usually only come in bright green.

Pteryphylla camellifolia, found as a nymph and fed leatherleaf viburnum, 2013

Relative of *Copiphora* in Los Tuxtlas, Mexico © Ondřej Hes

Stilpnochlora sp. male in Los Tuxtlas, Mexico © Ondřej Hes

Life Cycle: The female chews on a branch and lays around a hundred eggs in one shot. This only occurs once or twice for each female, which dies a month or two later. The eggs take about two months to hatch.

Husbandry: High humidity and good ventilation are recommended. They eat the common leaf fares of blackberry, rose, oak, etc. Cannibalism is not common if there is adequate space and food.

Metrioptera brachyptera, Belgium (CC-BY Frank Vassen)

Decticus verrucivorus
Belgium
(CC-BY Frank Vassen)

Winged predatory katydid
Clonia wahlbergi
South Africa
(CC-BY-SA Bernard Dupont)

Stick katydid
Phasmodes ranatriformis
Australia
(CC-BY Jean and Fred Hort)

Chabrier's marbled katydid
Eupholidoptera chabrieri, male
France
(CC-BY-SA Bernard Dupont)

Straight-lanced meadow katydid
Conocephalus strictus
Virginia
(CC-BY Judy Gallagher)

Eastern shield-backed katydid
Atlanticus sp.
Virginia
(CC-BY Judy Gallagher)

(Top and center) Katydid nymph
Possibly *Metaballus litus*
Australia
(CC-BY Jean and Fred Hort)

Handsome meadow katydid
Orchelimum pulchellum
Virginia
(CC-BY Judy Gallagher)

Ant-mimicry: katydid nymph
West Papua, Indonesia
(CC-BY-SA G. Bohne)

Typophyllum mortuifolium
Ecuador
(CC-BY-SA Pavel Kirillov)

Typophyllum sp.
Ecuador
(CC-BY-SA Pavel Kirillov)

Armored katydid, *Acanthoplus longipes*, South Africa (CC-BY-SA Bernard Dupont)

Armored katydid, *Acanthoplus* sp., South Africa (CC-BY-SA Bernard Dupont)

Lichen-mimicking katydid, *Trachyzulpha* sp., Malaysia (CC-BY-SA Pavel Kirillov)

Lichen-mimicking katydid, *Dysonia simplicipes,* Ecuador (CC-BY-SA Pavel Kirillov)

Acripeza reticulata is one of Australia's most iconic insect species, largely due to its spectacular threat display. The revelation of vivid colors beneath the animals' wings may startle predators and buy time for the katydid to escape.
© Jackson Nugent

Nymph stages of specimens reared in Australia, hatched from eggs laid by a wild-caught female, and fed on a leafy vine, *Parsonsia straminea*
© Jackson Nugent

These mountain katydid eggs were laid on the screen lid, but most were laid on the plant they fed on. The green katydid is *Torbia viridissima,* the gum-leaf katydid, also from Australia.
© Jackson Nugent

The cryptic adult female is difficult to see on bark.
© Jackson Nugent

The adult male mountain katydid is shaped like a normal katydid. It can fly and sing. The female is silent.
© Jackson Nugent

Mountain katydid hatchling
© Jackson Nugent

Predatory katydid on Tioman Island, Malaysia © Ondřej Hes

Tiny adult female predatory katydid on Tioman Island, Malaysia © Ondřej Hes

CLOSING

Captive breeding and laboratory husbandry parameters for various orthopterans are important to establish, not only to develop control methods for pests, or to cultivate and produce biocides. Moreover, it is the only possible key to the ultimate survival of many orthopteran species in the distant future. Habitat loss and change driven by human population growth is too intense for many habitats and their unique species. There are drawbacks to captive breeding, as it cannot save all species (very few in its present, limited form), is affected by human folly, and species in captivity change over time. However, none of these are good reasons to avoid this tool for saving species for future generations. The greatest risk may be the enthusiast themselves. Over time a percentage will research and discover the loss of habitat and problems caused by man and campaign against the problems. The answer of the politician is never to fix the problem, but always to shoot the messenger. Very likely their keeping of orthopterans will be seen as part of the problem—it is certainly the easiest of threats to squash with legislation that will follow. There are many laws against native and non-native wildlife husbandry across the world. Humans will not question the risk of life and limb for money and survival (the greatest cause of danger to natural habitats), but not for the simple joy of caring for and learning about creatures.

Hopefully, this book will inspire readers to study and consider orthopteran husbandry as a passion. Without passion we are all just robotic hordes of locusts cannibalizing without thought. Readers do not have to be zoo keepers or reside in countries without restrictive laws to study and enjoy orthopterans as pets. The variety of fantastic orthopterans in every back yard offers a lifetime of study and enjoyment.

Nymphs overwinter as purple or purple-and-green in this Brunswick, Ohio, population.

A handsome trig, *Phyllopalpus pulchellus*, Brunswick, Ohio, 2021

Slender meadow katydid
Conocephalus fasciatus
Brunswick, Ohio

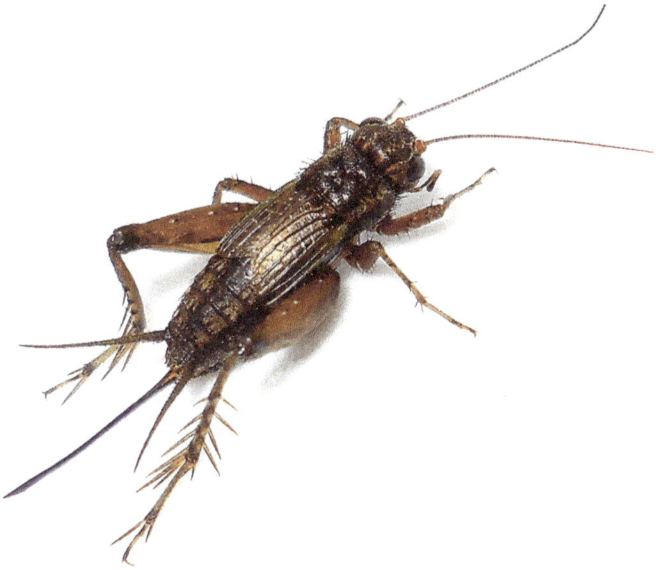

Striped ground cricket
Allonemobius fasciatus
Brunswick, Ohio
These number in the thousands in my tiny lawn in late summer.

Stethophyma grossum
Belgium

Grasshopper in Wuhan, China, Gunam Mountain © Ondřej Hes

Grasshopper in Wuhan, China, Gunam Mountain © Ondřej Hes

GLOSSARY

Abdomen: The last section of the insect body, attached to the thorax; contains the genital organs and cerci.

Adventive: Non-native species which has become established; naturalized.

Apterous: To lack wings.

Anterior: At or towards the front.

Apex: Point furthest from the point of attachment.

Autotomy: The reflexive release of an appendage at a specific regrowth point.

Brachypterous: short winged.

Cercus (*pl.* Cerci): Paired appendages at the end of the insect's abdomen.

Coxa (*pl.* Coxae): Basal leg segment.

Crypsis: Mimicry of common background objects like rocks or plants rather than fauna.

Deimatic display: Threat display or warning pose meant to discourage predators or competitors.

Entomophagous: Insect eating.

Desiccate: To dry out.

Diapause: A period of inactivity or arrested development influenced by seasonal changes in temperature or humidity, such as overwintering.

Dimorphic: Possessing two distinct forms or shapes. Sexually dimorphic males and females are visibly different in shape.

Dorsal: The upper surface or top.

Euplantula (*pl.* Euplantulae): Adhesive pad on the orthopteran tarsus used in climbing smooth surfaces; similar to pulvillus.

Femur (*pl.* Femora): Thigh; segment of an insect leg between the trochanter and tibia.

Frass: The solid waste of grasshoppers and other invertebrates.

Instar: Nymphal stages after hatching and between molts (i.e. a third instar nymph would have molted twice). Common abbreviation for instars is L1, L2, etc. The L comes from Larven-stadium, the German word for instar.

Margin: A border or edge.

Mesothorax: Middle section of the thorax; contains the tegmina and middle legs.

Metathorax: Terminal section of the thorax, attached to the abdomen; contains hind legs and hind wings.

Molt: The act of shedding the exoskeleton.

Nymph: Stage between egg and adult; for insects with an incomplete metamorphosis.

Ocellus (*pl.* Ocelli): Light detection organs.

Ommatidium (*pl.* Ommatidia): Structural unit or complete facet of the compound eye.

Bark mimic katydid in Kruger National Park, South Africa © Ondřej Hes

Small forest grasshopper on Mount Kinabalu in Borneo © Ondřej Hes

Ootheca (*pl.* Oothecae): A group of eggs produced together and encased by a foamy protein secretion; egg pod.

Ovipositor: Tubular organ through which an insect lays eggs.

Parthenogenetic: Of or having the ability to reproduce through the development of unfertilized eggs.

Pronotum: Shield-like upper surface of prothorax.

Prothorax: Front section of the thorax; attached to the head and contains front pair of legs.

Spermatheca: Female's sperm storage compartment.

Spermatophore: Sperm packet.

Spermatophylax: Gelatinous protein surrounding the spermatophore, when it is significant (as in most Orthoptera).

Spiracle: The external opening of the tracheal respiratory system; breathing hole.

Sternite: Ventral abdominal plates.

Stridulation: The act of rubbing body parts, often uniquely modified, together to make noises for attracting mates or warning predators.

Tarsus (*pl.* Tarsi): The foot of an insect. The tarsus contains a number of small segments including the claws.

Tegmen (*pl.* Tegmina): The tough and thickened front pair of wings.

Teneral: Soft, often pale or white colored state of the insect exoskeleton immediately following a molt.

Thorax: Middle section of the insect's body; in between the head and abdomen.

Tibia (*pl.* Tibiae): Segment of an insect leg between the tarsus and femur.

Trachea: Internal breathing tube that opens at a spiracle and branches out to the body cells.

Ventral: Bottom or underside.

Prognathogryllus sp., Waihee Ridge Trail, Maui, Hawaii (CC-BY Forest and Kim Starr)

Bladder grasshopper, *Bullacris unicolor,* male, South Africa (CC-BY-SA Bernard Dupont)

Large painted locust, *Schistocerca melanocera,* Galapagos (CC-BY NH53)

Monkey hopper, *Eumastax vittata,* Ecuador (CC-BY-SA Pavel Kirillov)

Monkey hopper, *Pseudomastax personata,* Ecuador (CC-BY-SA Pavel Kirillov)

Cornops sp.
Ecuador

Stropis sp.
Australia

Traulia sp.
Malaysia

BIBLIOGRAPHY

Ball, E. D., R. Tinkham, and C. T. Vorhies. 1942. The grasshoppers and other Orthoptera of Arizona. *Arizona Agricultural Experiment Station, Technical Bulletin* 93: 255-373.

Brock, Paul. 1993. Studies on stick-insects of the genus *Leptynia* in Spain. *AES Bulletin* 52(389).

Bruins, Eugene. 1999. *Encyclopedia of Terrarium*. The Netherlands: Rebo.

Brunner von Wattenwyl, Karl. 1890. Monographie der Proscopiden. *Verhandlungen der Kaiserlich-Königlichen Zoologisch-Botanischen Gesellschaft in Wien* 40: 87-124.

Capinera, J. L., and T. S. Sechrist. 1982. *Grasshoppers (Acrididae) of Colorado: Identification, Biology, and Management*. Colorado State University Experiment Station, Fort Collins Bulletin 584S.

Chang, Ester. 2014. Raising dragons: Managed care and breeding of the dragon-headed katydid (*Lesina intermedia*) at the San Diego Zoo. *Invertebrates in Education and Conservation Conference (Terrestrial Invertebrate Taxon Advisory Group).*

Clapper, Nikki Bruno. 2016. *Crickets*. North Mancato, Minnesota: Capstone.

Cockayne, E. A. 1930. Insect teratology, Part I. Homeosis and heteromorphosis and Part II. Extra wings. *Transactions of the Entomological Society of London* 78: 209-226.

Dunn, Gary A., ed. 1993. *Caring for Insect Livestock: An Insect Rearing Manual*. Lansing, Michigan: YES.

Fernandes, Maria, Adilson Zacaro, and Jose Serrao. 2015. First report of a parthenogenetic Grylloidea and new genus of Neoaclini (Insecta: Orthoptera: Grylloidea: Phalangopsidae: Phalangopsinae). *Zootaxa* 4032(4): 407-416.

Fernández, Miquel Domenech. 2019. Introducción a los ortópteros de España. *Mundo ArtróPodo* (No. 6; July): 5-14.

Froeschner, R. C. 1954. The grasshoppers and other Orthoptera of Iowa. *Iowa State Journal of Science* 29: 163-354.

Greven, Hartmut, Sabrina Braatz, and Dieter Shulter. 2013. Comments on the Malaysian Katydid *Ancylecha fenestrata* (Fabricius, 1793) (Orthoptera: Tettigoniidae). *Entomologie heute* 25: 57-75.

Haes, E. C. M. 1992. Crickets in Britain. *AES Bulletin* 51(380): 1-7.

Hes, Ondřej. 2005. To Croatian island Vir to visit not only the Orthoptera. *Sklípkan: Journal for Invertebrates Keeper and Breeders* 9(2): 74-78.

Hes, Ondřej. 2020. Keeping and breeding the horsehead grasshopper *Pseudoproscopia latirostris* (Brunner von Wattenwyl, 1890). *Invertebrates-Magazine* 19(3): 14-16.

Hromadka, J. 2000. Koylka sago pedo. *Sklipkan* '00(6): 134-135.

Kaur, Delvinder, Luis Carlos Neves, and Francis Cabana. 2018. Rearing unlikely dragons: Breeding of the dragon headed katydids, *Eumegalodon blanchardi* (Orthoptera: Tettigoniidae). *Journal of Entomology and Zoology Studies* 6(4): 734-739.

Idowu, A. B. 1997. The defensive mechanisms of *Zonocerus variegatus* (L.) (Orthoptera: Pyrogomorphidae) against potential predators. *Journal of African Zoology* 111(3): 199-203.

Kevan, D. K. 1953. A teratological specimen of *Taphronota calliparea* (Schaum). (Orthoptera: Acrididae) *Journal of the Entomological Society of Southern Africa* 16(2).

Key, R., and F. Key. 1936. Observations on rate of growth, coloration, and the abnormal six-instar life-cycle in *Locusta migratoria migratorioides*. *Bulletin of Entomological Research* 27(1): 77-85.

Krejci, R. 1999. *Tropidacris collaris*. *Sklipkan* '99(5): 102-107.

Larsen, Tom. 2002. *Aularches miliaris*—When your feelings get the better of you. [in'və:tibrət]: *The Danish Society of Keeping and Rearing Invertebrates in Captivity* 1(2): 10-15.

Larsen, Tom. 2002. A cricket that loves the darkness—*Phaeophilacris bredoides*. [in'və:tibrət]: *The Danish Society of Keeping and Rearing Invertebrates in Captivity* 1(3): 4-7.

Larsen, Tom. 2011a. *Ancylecha fenestrata*. *Exotiske Insekter* (No. 84): 42-48.

Larsen, Tom. 2011b. Graeshopper—nr. 3 *Zonocerus variegatus*. *Exotiske Insekter* (No. 88): 37-44.

Larsen, Tom. 2019. Cry of the suzumushi. *Invertebrates-Magazine* 19(2): 6-11.

Larsen, Tom. 2020. The world's largest (grasshopper). *Invertebrates-Magazine* 19(3): 2-7.

McMonigle, Orin. 2005. Featured Invertebrate: *Neobarrettia spinosa*. *Invertebrates-Magazine* 5(1): 16-17.

McMonigle, Orin. 2006. The grasshopper clown *Romalea microptera* (Beauvois). *Invertebrates-Magazine* 6(1): 6-10.

McMonigle, Orin. 2011. *Invertebrates for Exhibition: Insects, Arachnids, and Other Invertebrates Suitable for Display in Classrooms, Museums, and Insect Zoos*. Landisville, PA: Coachwhip.

McMonigle, Orin. 2013. *Keeping the Praying Mantis*. Greenville, Ohio: Coachwhip.

Mitchell, J. M., C. W. Stockton, and D. M. Meko. 1979. Evidence of 22-year rhythm of drought in the western United States related to the Hale Solar Cycle since the 17th century. pp. 125-143 from, *Solar-Terrestrial Influences on Weather and Climate*. Holland: Reidel Pub.

Preston-Mafham, Ken. 1992. *Grasshoppers and Mantids of the World*. Wellington House, London: Cassell.

Rainey, Reginald C. 1989. *Migration and Meteorology*. Oxford: Clarendon.

Rogez, Leon. 1996. *Tropidacris collaris:* Criquet geant de Guyane. *Terrario Magazine* (4): 15-18.

Simon, Seymour. 1975. *Pets in a Jar*: *Collecting and Caring for Small Wild Animals*. New York: Viking.

Stanek, V. J. 1969. *The Pictorial Encyclopedia of Insects*. London: Hamlyn.

Sutton, Peter G. 1992. The purple form of the wart-biter, *Decticus verrucivorus*. *AES Bulletin* 51(380).

Sutton, Peter G. 2004. Class entomological sites: Branscombe Beach and Undercliff, East Devon. *AES Bulletin* 63(453).

Weissman, David B., Thomas J. Walker, and David Gray. 2009. The field cricket *Gryllus assimilis* and two new sister species (Orthoptera: Gryllidae). *Annals of the Entomological Society of America* 102(3): 367-380.

Weissman, David B. 2021. Generic relationships of New World Jerusalem crickets (Orthoptera: Stenopelmatoidea: Stenopelmatinae), including all known species of *Stenopelmatus*. *Zootaxa* 4917(1): 1-122.

White, M. J. D. 1968. A gynandromorphic grasshopper produced by double fertilization. *Australian Journal of Zoology* 16(1): 101-109.

Amblycorpyha insolita, Santa Catalina Mountains, Arizona (CC-BY Katja Schulz)

ASSASSIN BUGS, WATERSCORPIONS, AND OTHER HEMIPTERA
REPRODUCTIVE BIOLOGY AND LABORATORY CULTURE METHODS
Orin McMonigle

CENTIPEDES IN CAPTIVITY
The Reproductive Biology and Husbandry of Chilopoda
Orin McMonigle

BREEDING THE WORLD'S LARGEST LIVING ARACHNID
AMBLYPYGID BIOLOGY, NATURAL HISTORY, AND CAPTIVE HUSBANDRY
ORIN MCMONIGLE

INVERTEBRATES FOR EXHIBITION
ORIN MCMONIGLE
INSECTS, ARACHNIDS, AND OTHER INVERTEBRATES SUITABLE FOR DISPLAY IN CLASSROOMS, MUSEUMS, AND INSECT ZOOS

MILLIPEDS IN CAPTIVITY
DIPLOPODAN HUSBANDRY AND
REPRODUCTIVE BIOLOGY
ORIN A. MCMONIGLE

FORGOTTEN
ORDER OF THE
VINEGAROONS
Orin McMonigle
WHIPSCORPION
BIOLOGY, HUSBANDRY,
AND NATURAL HISTORY

PILLBUGS AND
OTHER ISOPODS
CULTIVATING VIVARIUM CLEAN-UP CREWS AND
FEEDERS FOR DART FROGS, ARACHNIDS, AND INSECTS
Orin McMonigle

KEEPING THE
PRAYING MANTIS
Orin A. McMonigle

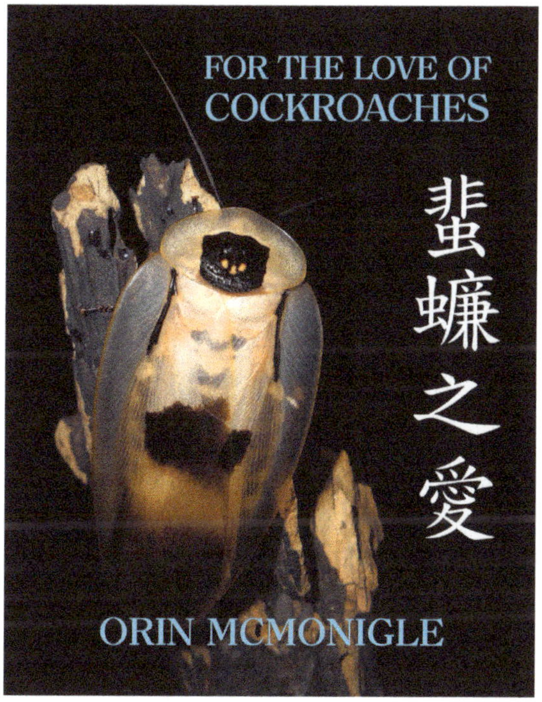
FOR THE LOVE OF
COCKROACHES
蜚蠊之愛
ORIN MCMONIGLE

THE ULTIMATE GUIDE
TO BREEDING BEETLES
COLEOPTERA LABORATORY CULTURE METHODS
ORIN MCMONIGLE

BREEDING THE VAMPIRE
AND OTHER CRABS
(BRACHYURA AND ANOMURA IN CAPTIVITY)
Husbandry, Reproduction,
Biology, and Diversity
ORIN MCMONIGLE

ORIN MCMONIGLE
ISOPOD
ZOOLOGY
Biology, Husbandry,
Species, and Cultivars

www.ingramcontent.com/pod-product-compliance
Ingram Content Group UK Ltd.
Pitfield, Milton Keynes, MK11 3LW, UK
UKHW061808111026
15625UKWH00044B/427